Walking Catfish & Other Aliens

*to Mother, Betty and Eric
for long years of faith and support*

Walking Catfish and Other Aliens

CHARLES E. ROTH

 Addison-Wesley

Also by Charles E. Roth
THE MOST DANGEROUS ANIMAL IN THE WORLD

 An Addisonian Press Book

Text Copyright © 1973 by Charles E. Roth
Illustrations Copyright © 1973 by Addison Wesley Publishing Co., Inc.
All Rights Reserved
Addison-Wesley Publishing Company, Inc.
Reading, Massachusetts 01867
Printed in the United States of America
Third Printing
WZ/WZ   5/75   06528

Library of Congress Cataloging in Publication Data

Roth, Charles Edmund, 1934-
   Walking catfish and other aliens.
   SUMMARY: Describes various animal species not native to the United States, how they arrived on the North American continent, and their effect on the native wildlife.
   "An Addisonian press book."
    1. Animal introduction—United States—Juvenile literature. [1. Animal introduction] I. Title.
QL86.R67   591.5   72-7436
ISBN 0-201-06528-2

Contents

	Introduction	7
	Alien Mammals	**11**
1	A Coat of Fur—the Coypu or Nutria	15
2	In Forest and On Farms—Wild Boars and Feral Hogs	20
3	Family Feud—Black and Norway Rats	29
4	After the West is Won—Burros and Horses	38
	Alien Birds	**55**
5	Feathered Vagabond—Cattle Egret	59
6	That Man May Hunt—the Ringnecked Pheasant	66
7	A Bird of Royalty in America—Mute Swan	74
8	They Turned the Tide—Starling and House Sparrow	82
	Alien Fishes	**97**
9	Introducing the Biggest Minnow—the Carp	99
10	The Fish that Runs Away—Walking Catfish	105
11	The Brook Trout Gains a Rival—Brown Trout	110
	Alien Insects	**117**
12	The Great Caterpillar War—the Gypsy Moth	124
13	With Sting of Fire—the Imported Fire Ant	138
14	Bringing in the Troops—Vedalias and Others	149
	Man and Other Aliens	**159**
	Acknowledgements	172
	Index	173

Introduction

The barking dog roused the curiosity of the night watchman. The bark was not the one it used for people. The watchman decided to investigate. He was not a drinking man and was completely unprepared for what he found. The dog was slowly backing away from, believe it or not, a large pink catfish that was walking toward it on spiny fins. The watchman could scarcely believe his eyes. He was seeing one of the first walking catfish found in Florida. How did it get there? Where did it come from?

There are many other animals such as the walking catfish which are relative newcomers to North America. One of these is man himself. This book explores the stories of a number of them. A few of the animals arrived here through their own wanderlust, such as man and the cattle egret. The rest came largely with the help of men. Some came directly as invited guests; many more came as stowaways on boats and planes.

These "new" Americans are full of interesting problems and possibilities for our native wildlife communities. Some of the exotic species have become welcomed additions to the American scene. Others have become outright pests and villains.

We are still inviting a variety of alien creatures to share the riches of this great land. But we must learn more about them. Only with sound information can we securely extend a firm invitation to take up life here. We should also learn how to guard against future villains. Such information is found on the following pages. Enjoy yourself.

C.R.
Littleton, Massachusetts
September, 1973

Walking Catfish
&
Other Aliens

ALIEN

MAMMALS

Humans seem to have a closer attraction to mammals than to most other animals. Perhaps it is because we are mammals ourselves. Also we have learned to tame many mammals to serve a variety of our needs for food, clothing and transportation. Others, such as dogs, cats and smaller pets, seem to serve our emotional psychological desires as much or more than practical necessities.

The American continent was blessed with a great wealth of mammals. Through the ages herds of mammoths and mastodons, bisons, caribou, pronghorns, elk, deer, and many other plant eaters abounded on the great plains and forests. Animals which preyed upon them (predators), such as the sabertoothed cat and dire wolf, the timber wolf and cougar, sought them out. While lesser predators, such as weasels, otters, bobcats, badgers and foxes, found food in smaller plant eating creatures: rabbits, prairie dogs, woodchucks, and a host of other rodents.

During several periods in the history of North America, a land bridge existed between North America and Asia in the vicinity of Alaska's Bering Straits. The horse and the camel that originated in North America crossed this bridge to a new home in Asia. Then they died out in America. On the other hand, some creatures came from Asia to America, including one species that was destined to alter the whole face of the continent. That species was man. According to the best evidence to date, he first wandered down the gaps between the great ice sheets more than 10,000 years ago.

The evidence, left by the earliest human visitors to America, is very spotty. Most scientists feel that these men brought no other animals with them except the fleas, lice and worms that were their parasites. Some suspect that the domesticated dog came with man.

Later invasions of peoples from Europe brought other animals with them to add to the fauna of North America. Of the mammals that came to America most were brought because they could serve man's needs. But some, such as the house mouse and its cousins the rats, came to take advantage of the bounty of human activities.

Once in a while man brought an animal to this country that already existed in small numbers. An example is the red fox. This fox is found all across the northern hemisphere but was very rare in the forests of eastern North America.

The English gentlemen who came to America wanted to have their traditional fox hunts. They found native gray foxes poor substitutes for the wily English red fox. So the Englishmen imported their own reds. As the forest was cleared for farms many foxes escaped and found suitable homes.

In time, the imported English foxes spread west into the territories of the native red foxes and the two mixed together. Today, scientists no longer consider the two foxes as separate species but only as one. Thus the red fox is both a native species and an invited guest.

In the following chapters we will explore the stories of several other mammals that have reached America's shores and found a new home.

ONE
A Coat of Fur

THE COYPU OR NUTRIA

The mid-1930's model truck bouncing over the country road had Louisiana license plates. In the back of the truck were several crates addressed to Mr. E. A. McIlhenny. They contained what McIlhenny hoped would be the beginning of a whole new fur ranching industry.

Within a short while McIlhenny was leaning on the frame of the wire enclosure carefully watching the activities of his new charges. He would experiment with these animals to see if he could find inexpensive and simple ways to raise large numbers of them in captivity. The animals he was watching had the large front gnawing teeth of the rodent clan. Their faces were blunt and squarish like guinea pigs' and porcupines' rather than pointed like most North American rodents'.

These imports from South America were dark brown in color and had the webbed feet of water dwelling mammals. They were larger than muskrats but smaller than beavers. If exotic names would help sell its fur this animal would be an instant success. It is called either coypu or nutria.

Nutria have long, coarse outside hairs that conceal the soft, velvety underfur that man uses.

McIlhenny was not the first to bring in these South American furbearers. In 1899, Will Frakes of California brought in a small colony but they soon died. Another fur rancher in Washington state tried in 1932, but he also was unsuccessful. McIlhenny had some success. He kept them in individual wire pens as well as great concrete tanks. These creatures could really eat. Feeding them kept him very busy, but his colony grew.

Early in his experiments, one of the unpredictable hurricanes whirled in off the Caribbean and slammed into the heart of Louisiana. In the confusion some of the nutrias panicked and dug their way to freedom. These were the first nutrias to escape to the best of our knowledge. Then in 1940, the Mississippi River went on one of its frequent rampages and flooded the fur ranch, smashing pens as it went. Hundreds of these aquatic animals were set afloat in the great marshes of the lower Mississippi. They found this new country much to their liking and set up housekeeping. The nutria became a new addition to the American melting pot.

The nutria is very large as rodents go, nearly three feet long. Fourteen inches of this is tail but it is still a big rodent. At first glance the nutria looks like a giant muskrat. It has the same general shape, large webbed hind feet and a scaly, scantily haired tail. The nutria's tail is round, however, not flattened sideways like a muskrat's.

Nutrias reproduce abundantly. The young are born between 120 and 150 days after mating, and there may be as many as nine in a litter. Well formed at birth, with their eyes open, nutrias can take to the water within 24 hours after they are born. Nutrias may have two or three litters a year and the young can begin adding to the population by the time they are six months old. Since adult females live about four years and the males may live as much as eight years, you can see, by using a little arithmetic, that the numbers of this animal can grow rather quickly in a suitable area.

Full grown nutria average around 20 pounds. There are not too many predators in the bayou country which can tackle an animal of that size. Only the increasingly uncommon otter and the alligator pose much of a threat to adults, although young nutria face more enemies such as mink, foxes and some snakes. The facts are that the nutria is expanding in numbers much faster than its enemies; even man, the trapper of this furbearing animal, cannot cut them down.

A wide variety of aquatic plants fall prey to the nutria's appetite. Unfortunately for man, the diet doesn't stop with aquatics. Sugar cane, rice, corn, cabbage, lettuce, peas and other crops are very much to their liking. Naturally this does not endear them to farmers.

In Texas some shrewd people thought they knew how to put the nutria's appetite to work for man. They imported the animal to control water lilies, arrowheads, cattails and other waterweeds considered to be undesirable in certain Texas lakes. The nutria really did the job. All vegetation was wiped out and the lakes were turned into muddy potholes. They no longer supported ducks or muskrats; in fact, they became unsuitable for the nutria as well. The people outsmarted themselves.

Besides a big appetite, nutrias also have aggressive dispositions. They are vicious fighters both among their own kind and with neighbors like the muskrat. And it should be mentioned that the muskrat is America's most valuable furbearing animal. Nutrias not only drive muskrats out of their territory, but also eat the same food and much more of it. Where nutrias become common, the muskrat population declines. The muskrat itself is no slouch of a fighter, but the nutria has it outweighed and outclassed.

From initial escape in Louisiana, the nutria has been spreading out in all directions. They have now been reported as far north as Wisconsin, Ohio, Pennsylvania and western Canada. The species is on the move. It can be expected in virtually all of our states and much of Canada. It is only a matter of time.

Like muskrats, nutrias dig burrows. Although they often weave grass nests on the spongy masses of southern marshes, they prefer to live and bear their young in earthen burrows. The entrance is usually below the surface of the water. Since the animals can remain submerged for at least seven minutes at a time, digging such a burrow is fairly easy for them. The burrows usually slant up to the surface so an air vent can be made. These burrows, which cave in easily, are a terrific nuisance in dikes, levees and irrigation ditches and cost mankind many dollars annually in repair work.

In its native South America—Chile, Argentina, Uruguay, Paraguay and Bolivia—the nutria became an important furbearer with the fur being used for coats and felt. Its numbers were kept in natural check by a variety of predators, parasites and diseases. The added pressure of human trapping put the population of nutria on the decline. Since their flesh is eaten in South America, the pressure on the nutria was even more intense. Argentina finally had to pass laws to protect the animal from being completely wiped out.

Hoping to cash in on a new fur, Americans and Europeans imported the creature to establish fur ranches. Like many potential get rich quick schemes, this proved costly. Breeding stock was offered at inflated prices but the fur brought only two to ten dollars a pelt. Its habits have proved damaging to the interests of man and native wildlife. It has become a pest, consuming the food of other creatures, destroying crops and ruining earthworks. It gives little in return.

TWO

In Forest and on Farms

WILD BOARS AND FERAL HOGS

Myakka River State Park is Florida's largest and wildest. Our family had sought it out because of the good chances of seeing wildlife there. With our camp finally set up and raccoon-proofed to the best of our ability, we wanted to explore for the creatures of the night. We decided to drive by car down the Park's Scenic Drive to see what creatures might show up in our headlights.

Shortly we saw two spots of light that were glowing animal eyes. It was a large sow, and she had a litter of babies with her. She turned and crossed the road followed by the little pigs. We began to count. One, two, three, four, five, six, seven, eight . . . fifteen, sixteen, seventeen, eighteen . . . twenty, twenty-one, twenty-two . . . this was getting absurd! No hog had that many young. Although we had a great deal of respect for a sow with pigs, it was time to get out of the car and carefully check what was going on. By this time another car had pulled up behind us to watch as well. They were laughing merrily. I decided to find out why.

"What's so funny?" I asked.

"Well you don't realize it but you were just held up."

"What do you mean? The only thing we have seen is an unbelievable number of pigs with that old sow."

"That's just it. There were only eight little pigs. They would cross in front of you, run down the ditch, cross behind you, circle in the woods, then cross in front of you again."

"You're kidding! Why would they do it?"

"They were panhandling. Even at their tender age they know that tourists often feed them from cars."

Well, we certainly had been played for fools, but then again the wild pigs didn't get anything to eat either.

It certainly is different to be held up in a park by panhandling pigs rather than panhandling bears. However, interesting as the hogs are, we had to wonder what effect they had on the park. They use their powerful snouts to turn over the soil in search of a variety of food from seeds to small animals. As we hiked about the park we found many places where herds of hogs had been feeding. These spots often several hundred feet in area would look like they had been plowed. One turned over area, where apparently several herds had been feeding regularly, was almost the size of a football field. Surely such activity must be changing the general character of the plant life.

Others also wondered, for during the latter part of the 1960's a scientist from Michigan State University had undertaken an ecological study of Myakka River State Park. His name was Leslie Gysel and his work provided some answers about the hogs and their place in the ecology of the park.

Gysel identified nine different plant communities in Myakka River Park. The hogs sought food in most of them at one time of the year or another, but they preferred the live oak plant community. There they dug for acorns and acorn weevils. Gysel found that the hogs rooted almost continuously in the surface soil layers. They dug to depths of from four to nine inches. Soil was usually completely turned over. Only the most deeply rooted plants with flexible stems could withstand the disturbance. This meant that many weedy plants such as greenbriar, poison ivy, and peppervine prospered. Cabbage palm and laurel oak also survived.

Hogs generally turned over patches of earth almost 20 feet square or in long strips. But Gysel found that they sometimes turned over areas from two to five acres in size. As a result of the exposure of bare soil, many kinds of small herbs and grasses were finding a home, thus altering plant communities.

Gysel called for further research on a number of things related to the hogs, and he suggested at the same time that their numbers had to be greatly reduced for several reasons.

A large hog population eats large numbers of acorns and thus reduces the available food for native wildlife such as deer, turkey, and squirrels.

By rooting up the acorns and oak seedlings, the hogs prevent the development of young live oaks. It is possible that, in time, the live oaks would disappear and be replaced largely by cabbage palm. This would mean quite a change of scenery in the park.

The park personnel have been trapping many of the hogs although leaving enough for the tourist to see. This in itself creates a problem because tourists foolishly try to feed the hogs like they foolishly feed the bears in parks. They fail to understand that an angry sow hog with pigs is almost as dangerous an opponent as a sow bear with cubs!

In earlier times the hog population might have been kept under control by some of the natural predators and parasites such as wolves and panthers, but wolves were exterminated in south Florida sixty years ago and panthers are quite scarce.

Screwworms, however, might have become the needed natural parasite, but it seemed the maggots of the screwworm fly were also a very serious pest of cattle.

Myakka River State Park is not the only place in the country with wild hogs. In many southern states hogs have been allowed to roam free and then periodically rounded up. Quite a few, of course, manage to escape the roundups. Feral hogs, that is, hogs escaped from domestication, are still found in many places and they create problems similar to those at Myakka River Park.

Not all wild hogs come from former domestic stock. Some are wild boars, the wild ancestors of domestic hogs. Wild boars have been hunted in Europe as a sport of noblemen for many centuries. American hunters also wanted a chance to hunt the wily boars. Wealthy sportsmen from several states imported them. Tennessee and New Hampshire have the most successful herds but wild boars are to be found in limited numbers in Arkansas,

Feral hogs, those escaped from domestication, can cause as much damage as wild boars.

Young wild boars are striped—a form of protective coloration that their domestic relatives lack.

Florida, California, Georgia, Missouri, North Carolina, Oregon and Texas. In almost all these states the boars have intermixed with the feral pigs of the area and it is hard to find purebred wild boars.

Most of the imports came from Russia and Germany where they were abundant. It is reported that those from Russia were larger and darker, while those from Germany were smaller and browner.

The purest population of wild boars in America is in New Hampshire. The original stock, a herd of 50, was brought to Corbin's Park in Sullivan County, New Hampshire in the early 1890's, by Austin Corbin. Corbin's Park was a private game preserve of 25,000 acres surrounded by a 9 to 12 foot high, heavy wire fence, some 30 miles long.

Today the boars just about hold their own there. The females have litters of three to five which is considerably less than the litters of feral hogs. Each year there is hunting, under regulated conditions, for the animals. In addition, the harsh winter conditions of the area help keep the boar population in the park at a rather stable level.

Soon after the boar herd was established in Corbin's Park, an unknown number found a break in the fence and escaped. Perhaps 25 or 30 animals descended from these escapees and still roam several of the rural towns in New Hampshire. These animals drift about following the most plentiful supplies of food. Their numbers have not varied greatly in more than fifty years.

In Tennessee the story was much different. Those that escaped from a game park there thrived and multiplied rapidly, especially in the Cherokee National Forest. They became very plentiful until reduced by an epidemic of cholera in 1932.

Where boars are encouraged for hunting, it is quite necessary to keep them fenced in because of their extreme potential for destroying crops. In an hour the rooting about of a herd of boars can destroy an entire field. There is always a good chance of animals' escaping these enclosures as happened in Tennessee and New Hampshire. In Tennessee the area where the boars escaped was largely mountainous with little agricultural activity. So they have caused little economic damage and are considered an important game animal. But in New Hampshire things are different. There are farm lands in the region. In 1949 the New Hampshire legislature passed an act requiring people responsible for introducing wild boars to capture or exterminate the animals and the offspring. The act also made them responsible for all boar-caused damage after April 1950. This law apparently hasn't had much effect on the hog population.

Boars and feral hogs are now among the fauna of several of our states. Just how welcome they should be is still a matter of debate. It is certain that they cause changes in plant communities due to their rooting about. Whether this is good or bad depends upon your point of view. In Germany a study of the Saupark of Springe, a famous wild boar forest, indicated that the animals were a tremendous economic asset. Acre for acre the boar forest was the most profitable in the region. The working of the forest floor and the removing of insects increased forest production.

In this country, studies show competition with other wildlife for food and changes in the makeup of some forest communities. Not enough work has been done to indicate whether or not these are beneficial changes to the region. Yet there is no doubt that feral hogs and wild boars can be very destructive in agricultural areas. All in all, hogs seem to be a mixed blessing as an addition to the mammals of America.

THREE

Family Feud

BLACK AND NORWAY RATS

As the ship sailed up the river, the men aboard could just begin to make out the tiny colony nestled against the dense forest. The fort was the most prominent feature, but cabins could be spotted nearby. To the sailors it surely seemed a wild and desolate spot—this Jamestown.

This colony of England had been established only two years before in 1607. Although the British had attempted a colony in America years earlier on Roanoke Island in Carolina, it had failed. Jamestown was surviving but it was no easy task. The colonists wrestled a living from the land by hunting and by simple agriculture. They raised corn, chickens, and some other livestock. It was a hard life and these hardy pioneers eagerly looked forward to the rare ships that sailed into Jamestown harbor bringing precious supplies and news from the motherland.

By the time the ship dropped anchor in midstream and the boats had been lowered over the side, most of the colonists were lined up on the shore cheering the sailors on. Eagerly they waded out to catch the bow of the boats as they came ashore. After the greetings, sailors and colonists joined forces to unload the ship's cargo and bring it ashore. When the ship left to return to England it would also carry a very special cargo—Captain John Smith. He was returning to England to recover from a serious injury.

Although they would miss their stricken leader, the isolated colonists welcomed most of the newly arrived cargo—except for one item—rats. These rats, probably the first to set foot on North America, were black rats, not the common Norway or brown rats of today's city and farm. Black rats were common in all the cities of Europe in those days, particularly the harbor towns.

They were excellent climbers and jumpers and frequently made their way aboard ships. They may have hidden among the cargo being unloaded or they may have gone overboard and swam to shore. Perhaps they came by both routes.

Undoubtedly these rats had gone aboard the ship in England and found adequate food to tide them over the long journey to America. Sailors were used to rats aboard ship and on occasion they even caught them and ate them. Nonetheless the colonists didn't want or need rats, particularly at that time in their history. Aside from spreading disease, the rats ate considerable food of all sorts and often fouled what foodstuffs they didn't eat.

The winter of 1609–10 was known in the colony as the terrible "starving time." Within six months, 90 percent of the colonists had died. History does not record what part the rats played in increasing the death toll but we can be fairly sure that they probably destroyed more than one person's grain and other's rations and probably spread disease. Some of the rats also died, but as with the humans, enough of them made it through to keep the colony going.

In June of 1610, a new governor, Lord Delaware, arrived at the colony bringing more men and supplies. This helped somewhat to ease the discouragement of the first settlers. Later more settlers arrived and with them undoubtedly came more black rats. Where men went rats soon followed. As the colonies of men spread along the Atlantic seaboard, so did colonies of black rats.

More than any other rat, the black is a parasite of human society. It prefers to live in man-made structures such as lofts, barns, stables, and ships. There it makes its nests and raises its many offspring. Man is a sloppy gatherer and storer of many kinds of food. And since rats are not fussy eaters, they can always find ample food supplies around the homes of men.

Black rats and brown, or Norway, rats are about the same length, around sixteen inches long. The black rat is, however, a smaller rat with a longer tail. It has larger ears and a more pointed, less arched snout than its brown cousin. It is also lighter and more agile.

It has a glossy black coat when in its prime with long, jet black, or bluish-black, outside or guard hairs and grayish underfur. The underparts are lighter gray or silvery. This rat's coat can be rather soft and in the Middle Ages, it was common practice to use black rat pelts for inexpensive fur coats or trim.

However, rat coats were made only because rats were so common. When rats were that common, their parasites, particularly fleas, were numerous. They carried a variety of human diseases, including plague. Plague is 50 to 90 percent fatal to man. It has been estimated that in the first 1500 years of the Christian era there were 109 rat-caused disease epidemics, and from 1500 A.D. to 1720 A.D. there were 45 such epidemics.

The black rat can produce three to seven litters a year and averages five to seven young in each litter. Sometimes a litter may contain as many as a dozen. Even though the young are naked, blind and completely helpless at birth, it is only three months before they are producing litters of their own.

Since black rats prefer the company of man, or at least the places he lives and works, they have few enemies except man and his associates—the dog and the cat. There is one major exception which we will discuss a bit later.

All in all, rat populations multiplied in seaport cities like Boston, New York, Philadelphia, and New Orleans. However there were historical rumblings afoot in the thirteen colonies that were to change the destiny of the world for both humans and black rats. The seeds of revolution were sown. And in 1775 British troops and supply ships were on their way to America. With them came more rats as stow-a-ways. This time it was the Norway, or brown, rat. Norway rats had come to the Mediterranean countries from Asia sometime during the 16th century. They spread rapidly to England by ship around 1728.

Norway rats did not regularly seek out ships like their black cousin, but wherever grain was transported brown rats went. Thus the species found themselves on the ships carrying grain to supply the horses of the cavalry. The

British may not have been able to subdue their rebellious colonies, but Norway rats were here to stay. In turn, these small brown rodents would subdue their cousins the black rats.

We stated earlier that black rats had relatively few natural enemies. That is true. However, one of their worst enemies is the Norway rat. They will kill and eat both young and adult black rats. By 1900, black rats had become rare in this country everywhere but along the Gulf Coast. Even there they had to continually avoid their more aggressive cousin.

Trading brown rats for black ones was no gain. In fact it was a distinct loss. Whereas black rats confined their activities largely to the buildings of man, Norway rats were more adventuresome. In winter they wisely sought the warmth and food of buildings, cellars and stables. But in the warmer months they generally moved out into the fields, sewers, and stream banks where they dug extensive burrows. Here they found safety and raised their young. Of course out in the fields, Norway rats must face predators that their black cousin seldom faced, such as hawks, owls, weasels, minks, and snakes.

However Norway rats are just as prolific as their black cousin, perhaps even more so. Fortunately the maximum life span for wild Norway rats is about two years. Even at that a female and her reproducing offspring can quickly build a very respectable population.

Norway rats have larger bodies and shorter tails than black rats, and their brownish gray fur is much coarser. Their undersides are dingy gray. They are certainly not attractive animals, even less attractive than black rats which can hardly be prized for beauty.

However, their habits are what make this animal among the most unlikeable of animals. Norway rats have vast appetites. They are often cannibalistic when their population becomes large or if other food is not available. Often they kill for sheer lust. They will enter a poultry house and kill every hen and chick in the place. Or they might eat the eggs. Even baby lambs and pigs are attacked. Norway rats destroy many crops in field and garden as well as stored fruits and vegetables. They have been known to bite and try to eat human babies in their cribs and even to attack sleeping adults. In zoos, elephants have even had their toes chewed off by rats.

Although this despised species has spread all across the nation, individual rats do not wander very far in the course of their activities. Scientists estimate that a 100-150 foot territory is an average "home range" for this creature. Within this limited area a rat will live and die, and it will fight fiercely to protect its territory. Any strange rat that appears is driven off or killed. Young rats are expected to find their own territories when they are grown up. These young adults are the travelers. They are the ones most likely to accidentally hitch rides of various kinds and move long distances. Actually it took rats 148 years to conquer this nation. Montana, the last rat free state, fell in 1923.

The Norway rat is with us to stay despite extensive efforts to poison him, trap him and ratproof buildings. It is a successful conqueror because it reproduces so successfully. Food sources are easy to find for he merely follows

Norway, or brown, rats destroy tons of grain and also seek out meat such as bird eggs and baby mammals.

man. This creature can dig under, swim around, or climb over most natural obstacles.

The Norway rat is one of man's most dangerous enemies. Not just because of its eating habits or the way in which it gnaws through materials, but because of its capacity to pass on diseases. The lice and fleas that make their homes upon the rat harbor some of our most dangerous illnesses: plague, typhus, infectious jaundice, trichinosis, rabies, and salmonella. There is also the very serious rat bite fever. It is hard to overestimate the danger of rats to people. We need to protect our health and welfare and see that all necessary steps are taken to control these pests.

The war on rats is a tough one. It promises to be long and drawn out. Intelligence reports suggest that there are 50 million rats on farms, 30 million in our towns, and 20 million more in the cities. Fighting rats is a job for all Americans wherever they live. The damage rats do costs every one of us about ten dollars a year.

To fight rats successfully we must remove all shelters they could use. This means keeping all areas free of every kind of litter. All food supplies must be placed in ratproof containers. All garbage must be kept tightly covered. And of course, the animals themselves must be destroyed. However killing rats is not enough. If the other things are not done an area cleared of rats soon becomes reinfested from surrounding areas.

The Norway rat's habit of moving out into the fields in the summer poses threats to native wildlife, particularly ground nesting birds. In Massachusetts, efforts to increase the nesting success of terns were seriously hampered by rats living on the sandspit the terns called home. Once the rats were brought under control the terns had a better chance.

On Martha's Vineyard, a small island off the Massachusetts coast, the heath hen made its last, and unsuccessful, stand against extinction. Rats were a serious predator on the bird's nests and hastened the death of that species.

The rat can adapt to a wide variety of conditions, and especially to a filthy environment. Indeed it seems to encompass all of man's worst characteristics. Perhaps that is one reason we hate the rat so much.

It came as a stowaway. It thrives. It pushes aside those that stand in its way. It is highly successful. It threatens wildlife; that is serious. It threatens man; it must go!

FOUR

After the West is Won

BURROS AND HORSES

Old Sourdough hadn't been as excited in years. Fifteen summers and winters had found him roaming these California hills always searching for the big payoff. A bit of dust; a nugget here and there kept him in grub. But now he had found his dream.

A vein of quartz near the top of the barren hill had caught his attention several days back. He and Whiskey Jack, the burro that carried his equipment, set up camp at the base. Each morning Sourdough hiked to the vein and slowly picked away at the rock until he had cut into the hillside more than the length of his body. Flecks of gold in the quartz kept the old man working from dawn to dusk.

At the end of the previous day's work the gold flecks had become very numerous. This day promised to bring the weary prospector face to face with his long sought fortune. He quickly went about his camp chores and when finished he tied the hobbles around Whiskey Jack's forelegs and turned him loose for his daily browsing on the desert shrubbery. Then he scrambled quickly up to his mine.

Sourdough dug as fast as his sinewy, but aging, muscles would permit. Meanwhile, Whiskey Jack moved slowly along the lower slope snatching a bit of grass wherever he found it or snapping off the tips of the creosote bush or bush lupine. Near midday the alert burro suddenly stopped his feeding. A low rumble and the bouncing of stones down the slope had caught his attention. He listened but nothing followed and he returned to his grazing.

That night Sourdough did not come down to camp. Nor would he. He had truly met his fate, a quartz lined grave on a nameless rocky hill.

Whiskey Jack missed his companion but hobbled about searching for food. Soon he became thirsty. The hobbles made moving about rather difficult. As thirst took firm hold of the creature, he fought the hobbles more and more.

Fortunately the ropes gave way after two days of battling. Within a few hours he had located water and was free and wild.

The clouds were piling up in threatening columns. On the plateau below, the two sheep herders were pushing their flock as fast as possible. They hoped to get near the water hole before the storm broke.

The bleating, swirling flock moved on, held intact by the men on horseback and the circling sheep dogs. Behind the flock, eating their dust, was a young man and five burros that carried the supplies for the herders.

The air was strangely quiet. The dust stirred by the sheep hooves settled quickly without the wind to push it around. The sheep were nervous and charged off in small bunches at the slightest strange event they encountered. The herders worked hard and constantly to keep the flock together. Even the burros were spooky and the young herdsmen were on the alert to prevent the loss of vital supplies they carried.

The flock had covered several miles since the midday break. Suddenly the wind came whipping in from behind and the men had all they could do to hold the sheep. Moments later the thunder clapped and lightning instantaneously shattered the air and struck the ground only several hundred yards away. The animals broke into a dead run—a true stampede. The burros joined. When the storm had passed, the herders were left with hours of rounding up their flock. Three burros, all females, managed to avoid the sheepmen and one carrying a pack soon scraped it loose on a rocky outcrop.

For several days the three burros browsed the nearby foothills. Then one day as they moved down to a new water hole, they came upon a lone burro—a male. He lost no time in galloping up to the newly arrived threesome and within a short time he had his harem. In time he would father a number of young that would increase this tiny herd.

These two stories are purely fictional, but they suggest some of the many ways in which burros gained their freedom in the American Southwest during the 1800's. Prospectors, sheepherders and other desert travelers were abundant and the hardy burro was their favorite beast of burden. The land was harsh and dangerous for human travelers. Often they found it beyond their endurance so they perished or moved many times. Their abandoned burros survived and found new homes.

Small herds of burros appeared in many of the states of the Southwest—California, Nevada, Arizona, New Mexico. Burros had been in this country since the Spanish came but it was not until the prospectors and sheepherders invaded the area in numbers in the 1800's that enough of the animals found their way to freedom and established wild herds.

The domestic burro originated from wild asses living on the great desert plains of North Africa. Their homeland provided sparse food and water. They survived well there and this made them valuable as beasts of burden in similar arid regions of the world. They were a natural for travel in the American deserts for they could easily live off the land.

Released from the service of man, feral burros found conditions in America to their liking. Food plants were abundant. Enemies were not common. Mountain lions were about the only predators that could kill a full grown burro and they were never common in the arid lands. Furthermore the sheepherders and cattlemen killed every mountain lion they could find. In fact, human hunters and disease were the main enemies the burros faced.

The number of burros grew steadily even though they were hunted for their meat. In later years they were even slaughtered for use in dog and cat food. Although burros usually have only one colt a year, the herds grow well. They have a long breeding season and about half the jennies (females) in a herd will bear colts in a given year. Burros live a long time and a jenny may bear as many as thirty colts in her lifetime. With so few natural enemies to remove animals from the herd, plus a good food supply, burro herds survived and grew.

With their round furry features and long floppy ears, these animals are very appealing to people. So when they were killed for pet food, many people were angered. It seemed a rather senseless slaughter. So in 1939, Californians passed a law prohibiting the killing of burros for the pet food industry. This was a start toward burro protection but it didn't stop individual hunting. Some people truly enjoyed the meat and hunted burros for that purpose. Other people just liked to kill. They would shoot the animals and leave them to rot where they fell. Since burros were not native game animals they were not protected by any laws but the one above. They could be hunted at any season.

The general public in California was now aroused by this wide open hunting and passed a law in 1953 prohibiting the killing of any wild burros for a two year period. Two years later they renewed the law for another two year period. This time the law went a bit further, however. It prohibited wounding, capturing or possessing wild burros. A provision was made by the state which allowed citizens twelve permits a year to capture a burro for a pet or a beast of burden.

The year 1957 was a banner one for burros in California. The law was made permanent. The only real changes stated that more burros could be captured under permit

It may take more than 100 years to reestablish the native desert plant communities destroyed by wild burros.

and landowners suffering proven damage to their property could get a permit to kill. In addition, a burro sanctuary was established in southeastern Inyo County.

It sounds like the fitting end of a tale of an animal threatened by man. It comes out the way we like. An appealing animal is saved from ruthless and thoughtless men. It might be nice if the story were that simple.

When the burro herds finally were relieved, even partially, from the hunting pressure of their chief enemy, man, the size of the herds increased. Further laws protecting them helped the herds grow even larger. By 1961 biologists estimated that the total number of burros in California alone ran between 2700 and 3700 animals. To burros and human burro sympathizers this was great. But there is still another side of the story.

Burros are very efficient foragers and can find food under very poor conditions. They prefer grass and weedy plants when they can get them but a large part of their diet is made up of browse—the more tender tips of shrubby plants or trees. The plants they prefer, such as California buckwheat, black bush, bush lupine, phlox and desert thorn, are the same ones that native wildlife depend upon for survival. Bighorn sheep, quail, burrowing mammals, small birds and even reptiles are affected by the burros' activities. In some places high populations of these animals have very seriously depleted, or completely destroyed, the range. They have destroyed it not only for other creatures but for themselves as well.

In a 1961 mimeographed report, a biologist from the California Department of Fish and Game reported:

"In Arcane Meadows on top of the Panamints (Mountains) near Telescope Peak, burros have nearly killed out phlox, a favorite tall flowering plant. Phlox now exists only in centers of large shrubs where burros cannot get at it. They are heavily browsing bush lupine and are uprooting much of it. The grasses in the meadow have been nearly killed out through close grazing and pawing.

"On the lower western slopes of the Panamints and in Butte Valley, the better browses have been killed out, and low value plants such as white bursage are nearly gone. The burros are now heavily eating the creosote bush. The foliage of this plant is rarely taken by any animal. Consumption of creosote bush reflects a very seriously depleted range condition. Old timers tell of former high numbers of bighorn sheep in the Panamints. At present a very low population exists. In the summer of 1955, during a partial survey of the western slopes and main ridge of the Panamints for wildlife species, more than 25 burros, one bighorn and fresh tracks of three other bighorn were observed. From examination of the serious damage wrought to the range by burros, it seems safe to assume that they have been the chief cause of the large reduction in bighorn numbers in this area."

Burro competition with desert bighorn sheep is a serious problem. The desert bighorn is one of our endangered wildlife species. It faces many problems. Disease has caused heavy losses. Until recently hunters seeking trophy horns had taken their toll. At this late date in bighorn history, burros may be one of the greatest continuing threats to their existence. This is true not only in California but in bighorn country throughout the Southwest.

Not only do burros destroy the food supply of these animals, but they ruin the precious water holes. Burros need more water than native desert wildlife and they tend to concentrate around springs and water holes. A burro herd may use up all the water at a small spring leaving little or nothing for bighorns or other wildlife. Furthermore, the trampling of their hooves compacts the soil and often shuts off the underground water arteries that provide water to the spring. In time the spring dries up. Such decreases in water and food spell serious trouble for the endangered bighorn.

In addition, not only do burros drink the water, but they eat up most of the vegetation around the water holes. This leaves little or nothing in the way of nesting, roosting and protection cover for a host of small birds and mammals. Normally the area round water holes and springs is the best habitat for such creatures.

In some places men have tried to improve desert watering places by piping water into concrete tanks. Underground water storage for game birds, called "quail guzzlers", are often built by game managers. In a number of cases burros have torn up the pipes, smashed the concrete and trampled out the guzzlers. Sometimes it is deliberate if the burros are frustrated when they can't get water. Sometimes it is due to accidents caused by males fighting for the attention of females. In any event the results have serious effects upon native wildlife which also need the water.

Actually it is through their total effect upon the fragile interacting system of desert life that burros became very unwelcome guests on the American desert. Their destruction of plants also robs the desert of its limited water holding power. Removal of plants robs desert rodents of food and shelter, and it is the burrowing activities of these animals that turn over the soil and spread it about. This provides more places for plants and thus more food and shelter for more birds, mammals and reptiles. And in turn larger predatory animals of the desert depend on the smaller creatures for food. Burro activity affects the whole web of life in the desert. To repair the damage they cause to the delicate, slowly changing desert takes long periods of time.

The burro, like many introduced species, has become too successful for his own good. Because of the appealing nature of his looks and the many tender stories of burros in children's literature, the burro has gained protection in some places. Under protection the animal thrives and ends up destroying the land for other creatures and even itself. In most areas, particularly in fully protected National parks and wildlife refuges, the burro will have to be severely controlled or eliminated if the desert habitat and its wildlife are to survive.

Perhaps more than any other creature, the future of the vanishing desert bighorn is tied to the way we humans handle the burro problem. I remember the hot, dusty, day-long trip under the hot sun from Tucson, Arizona to

Organ Pipe National Monument—one of the remaining homes of the bighorn. Early the next morning we were up and bouncing over dirt roads to get near spots where bighorns were known to frequent. For two days we searched. We saw burros, but no bighorns, in the vicinity of each of the water holes. On the second morning we did find one tiny water hole with the tracks of many sheep. Here there were no burros. Unfortunately, we never did see the makers of the sheep tracks and had to leave the area disappointed.

If the desert bighorn is to survive very far into the next century, if the desert is to maintain its variety of unique plants and the wildlife they support, burro numbers must be controlled. There are many who are as happy to see a burro as a bighorn. For these people there probably should be some sanctuaries for wild burros. Even in such places the numbers of burros will have to be kept within the ability of the habitat to feed and water them for their own health and survival.

In accidentally bringing the burro to our desert, man has tampered with a natural system. To protect that system man will have to deliberately step in and erase or reduce his error.

EL CABALLO—THE BURRO'S BIG COUSIN COMES TO TOWN

The burro's larger cousin, the horse, arrived in the New World around the same time as the burro but achieved wild status much earlier. Actually, the horse originated in

North America and spread from there to other parts of the world, most likely across a land bridge from Alaska to Asia. Then for presently unknown reasons wild horses became extinct in North America about 8,000 years before Columbus arrived. His arrival reintroduced the horse back into North America.

The Spanish explorers that came after Columbus brought their domesticated horses with them. It was no easy task. Their small, clumsy sailboats were very slow. The horses had to be kept in slings on the decks of these boats for as many as three or four months at a time. A great many of them died from lack of exercise, drenching storms, and constant exposure to the fierce, hot sun.

In certain areas of the route to America the winds often didn't blow for days on end and the boats stayed still. The scarce water would run low and horses would be killed and tossed overboard. The sight of the bloated dead horses floating in the water caused the sailors to name this area of little wind the "horse latitudes."

Most of the early Spanish visits were to islands in the Caribbean. Here they established horse ranches. Many fine horses were raised. Most were from desert Arab stock mixed with the Barb breed of horses brought to Spain by the Moors. This was to be important for horses from more northern climates probably would not have survived.

Even the trip from the islands to mainland North America was hazardous for horses. It was itself a three week voyage in those days.

When the famous Hernando Cortez landed in Vera Cruz on his invasion of Mexico, however, he had not only the original five mares and 11 stallions that he had left with but a foal had been born on board the ship.

The horses meant survival to the adventurous Spaniards. Horse meat kept them alive when they were lost and starving, and they found water when the men couldn't. They carried the sick and wounded. And at least in the beginning, they terrified the Indians.

It has often been suggested that wild horses escaped from the Spanish explorers. History does not support this theory. The horses were too valuable to the Spanish to let them escape. If any did escape they would be too few in number to start lasting herds. Some have suggested that the first wild horses were strays from Coronado's 1541 expedition to New Mexico. Again the records deny it; they show that all three mares on the roster died. Others have suggested the horses came from DeSoto's 1542 exploration of the Mississippi, but the official Spanish records reveal that the remaining eight horses of that ill-fated expedition were shot by Indians.

Actually, wild horses did not come until somewhat later. Indians were riding horses quite awhile before there were any wild horses on the prairies. In fact the domestic horses of these Indians were one of the sources of the great wild herds. Although the early Indians of Mexico originally feared the "man-horse" of the Spaniards, they soon learned that there was no magic. Plains Indians, however, did not show the same fears. And the Spanish sometimes bought their friendship with horses.

Often the horse changed the whole life style of these native peoples. For example, the Sioux were originally from the forests near the headwaters of the Mississippi. But they were pushed out of their land by the Ojibway or Chippewas. These defeated and now homeless people acquired horses from the Spanish. Within a short time they became a nation on horseback and were greatly feared by all other tribes. They had the most dreaded cavalry on the American plains.

The wild herds began around 1598, soon after the Spaniards settled in New Mexico and started large horse ranches. It was very difficult to keep close watch on cattle and horses in the large expanses of wild unfenced country that existed then. Often both cattle and horses strayed off and became free-living and wild.

Other historical events helped create the wild herds. The Pueblo villagers along New Mexico's Rio Grande didn't need or use horses but they did help create herds of wild mustangs. In 1680 they staged a successful revolt against the Spanish. In fact, they completely drove the Spanish out of New Mexico for twelve years. Hundreds of horses from the abandoned ranches became free and wild. The mustang herds had plenty of time to grow and spread out across the land.

Some of the herds came directly from Indians. The Apache and Navajo raided the ranches in New Mexico to gain horses. They even kidnapped Spanish stable hands to teach them how to ride. Both these tribes and others on the northern prairies lived a nomadic life. Their horse herds were always on the move with them. Under such conditions it was easy for horses to become lost, abandoned, or to escape.

Once wild herds were abundant, the Indians worried little about horse herding. It was easy enough to capture wild horses when necessary. Consequently, they never developed the art of horse breeding as the Spaniards had done in New Mexico and in Latin America.

The wild horses thrived in America. After all it had been the horse's ancestral home. American trappers and explorers moving west in the 1800's found hundreds of thousands of wild horses on the Plains. The individual herds themselves were often very large. Some early accounts report herds of as many as 40,000. One explorer reported there were so many horses between the Columbia River and the high desert country that "a single band traveled from dawn to dusk in passing a given point."

Today the wild horse is again headed for extinction in America. It is being driven out largely by descendants of the European men who brought the horse back to its native home. The plow, the cattle ranch and industrialization have cut up the range and deprived the horse of a home. The demand for horsemeat to feed millions of dogs and cats has deprived many of the horses of their lives.

In 1865, more than two million mustangs ranged from the Columbia River to the mouth of the Rio Grande. There were still somewhere around one million as late as 1925. Today a mere remnant, 10,000 to 20,000 wild horses survive in the American West. Their bands are tiny, ranging from 6 to 60. They are confined mainly to poor habitats not used by man for his activities. These wild horses are accused of competing with domestic livestock for food on the open range and the stallions of running away with domestic mares if they have a chance. They have few supporters among those living in horse country.

Nevada, Wyoming, Arizona, Utah, eastern Washington, Oregon, western Colorado and Montana have a few wild horses today. They are classified as domestic animals gone wild. Many are still trapped and sold for pet food. Others are shot because they compete with livestock or are caught particularly for rodeos. Unless actions are taken to give wild horses some protection they will probably be entirely gone by the year 2000.

There is some hope. Largely through the efforts of Mrs. Velma Johnston, affectionately known as "Wild Horse Annie," a 33,680 acre sanctuary for wild horses was set up on the Montana-Wyoming border. It is known as Pryor Mountain Wild Horse Range. It is home to around 200 horses. A second refuge was established in Nevada in 1962 on a training area for Nellis Air Force Base. Both areas are under the supervision of the Bureau of Land Management.

The legal battles are largely being fought by the International Society for the Protection of Mustangs and Burros under the leadership of "Wild Horse Annie" Johnston. This organization is particularly seeking legislation for more horse refuges on public lands.

Wild horses are beautiful creatures. The West would somehow not be complete without their thundering hooves.

Perhaps we owe it to ourselves and to history to see that some wild horses are a part of the future as well as the past.

ALIEN

BIRDS

Birds have fascinated men for ages. They are brightly colored, produce many beautiful sounds, and the eggs and meat of many kinds provide men with food. Furthermore, birds are primarily creatures of the day and are therefore active during the times man is.

Often these creatures have figured prominently in the music, art, and drama of many nations. As men have left to explore and settle new lands, they frequently felt a spot of loneliness for familiar birds they hunted in other days. When time and money permitted some tried to bring these old favorites into their new country. As you will see, this decision was not always beneficial to the land or to the people.

In America the heyday of bird importing came in the middle of the 19th century, although many imports were tried well before that time. There was a particular emphasis in establishing songbirds. Several societies were formed for this purpose. For example, they wanted the "robin redbreast," the English robin, imported. This bird is very different from our American robin. Such singers as the nightingale and the skylark were also greatly desired. Birds of Shakespeare were also sought; song thrushes, starlings, house sparrows and chaffinches. All of these were birds of Europe. Few of them, except the starling and house sparrow, found a real home in the New World. The English robins just wouldn't breed here. Nightingales usually arrived in poor health and died shortly after being released. Over a thousand skylarks were set free from the Atlantic to the Pacific. In some places small breeding flocks persisted for a number of years. In time most of these died out. Only in Victoria, British Columbia can these birds still be found wild and breeding in North America.

One cannot help but wonder what makes the Victoria region so special for it is here also that an Asiatic bird, the crested myna, has established a toehold on the continent. These birds can cause considerable agricultural damage and can be unpleasantly noisy. We are undoubtedly lucky that the myna has not bred elsewhere.

Of at least 44 different kinds of songbirds brought here only three species—the starling, the house sparrow, and the pigeon—have found conditions over the whole continent to their liking. Some species found homes in one small area.

The European tree sparrow is found in the vicinity of St. Louis, Missouri, and Springfield, Illinois, while the spot-breasted oriole, the red-whiskered bulbul, and the blue-gray tanager like the climate in Miami, Florida.

In addition to these imported birds many more species arrive as cage birds. A variety of finches including the familiar canary come to America each year to satisfy the pet trade. Many escape, some are deliberately set free. For almost all, such liberty has meant certain and quick death. None have become established as wild birds.

The American hunter has been responsible for successfully establishing more foreign birds than those who imported songbirds. As far back as Ben Franklin's day, wealthy landowners tried bringing in the ringnecked pheasant from England. The bird did not become a part of our landscape until much later. The ringnecked pheasant wasn't the only game bird to be invited here. The gray partridge, the chukar, the Coturnix quail, the francolin and many others have been tried. Most of these early attempts ended in failure.

Following World War II many former soldiers wanted to bring in game birds they had learned to hunt in other parts of the world. This posed a considerable problem, because many of these were potential pests as well as potential game species. To control the importing and protect the rights of all citizens, joint federal and state cooperation was needed.

This was achieved with the aid of the Wildlife Management Institute and in 1948 a foreign game program was begun. A major part of the program was to discourage unwise introductions. A great deal of ecological study was carried out both here and around the world. This country was surveyed to find what areas lacked native game species. Then the world was searched for species that could survive in the climate and habitat of these places, yet would not drive out native species or eat our crops.

Since 1950, more than 100 European and Asian birds have been invited to share the wealth in America. Of these, 14 species and four subspecies have received special attention. Only six of the 14 show much hope of becoming a permanent part of America's complex bird life.

FIVE
Feathered Vagabond
CATTLE EGRET

The weary white bird circled briefly and landed in a marshy field. It walked through the grass stirring up insects which it quickly grabbed to fill its empty stomach. The area where it had chosen to land was the broad floodplain of the Sudbury River. This slow moving stream is in the historic town of Concord, Massachusetts.

This was a strange bird for the Sudbury valley. It looked a great deal like an egret. This snowy white heron-like bird had been the early symbol of the bird protection movement and still graces the emblems of the National Audubon Society. However, it was different. It was stockier. The bill was stouter and shorter. Furthermore there was a yellowish color on the neck and back. The bird also stood more erectly than a snowy egret would.

All these characteristics were not lost on the three men who stood watching the bird through their binoculars on April 23, 1952. Massachusetts probably has more amateur birdwatchers than any other state, but these three, Dr. William H. Drury, Jr., Allen Morgan, and Richard Stackpole, were more serious ornithologists.

Although they were not professionals at that time in their careers, all of them realized that they probably were looking at a Massachusetts' first. Everything they saw added up to cattle egret. But this was an Old World bird. There were a few in South America but to the best of their knowledge there were none in North America. Was it an escapee from a zoo? Not likely. Perhaps it was blown in on a storm. Possible, but unlikely this far north.

The questions were many. Unfortunately no one would believe they had seen a cattle egret. There was only one thing to do. According to the cold code of science, the bird would have to be collected as positive proof. These men did possess scientific collecting permits, so a shotgun was obtained and the unfortunate bird was shot. Today it exists as a study skin in the collection of Harvard University's Museum of Comparative Zoology.

At first it was assumed that the bird was just a lone wanderer that had strayed far off its course to become the first of its kind to visit Massachusetts. However, during the summer Drury, Morgan, and Stackpole were viewing some film made by their mutual friend Dick Borden. Borden was an official photographer for the Boston Red Sox, but took every spare moment to photograph birds and other wildlife. On March 12th of 1952, he had seen some snowy egrets feeding among cattle at Eagle Bay Ranch near Okeechobee City, Florida. The sight interested him so he had stopped to photograph it. When Morgan, Stackpole and Drury saw the film they recognized immediately that the birds were not snowy egrets at all, but cattle egrets like the one they had collected in the Sudbury Valley.

As the news of the presence of cattle egrets in the United States was released, new reports began to come in from Chicago and New Jersey. Thus 1952 gave witness to a truly unprecedented phenomenon, the introduction of a foreign bird without the help of man. They came not as one or two isolated stragglers. They came in numbers. To date no one knows exactly why. They were probably not the first cattle egrets to arrive on this continent, however. In May of 1948 a staff member of the Everglades National Park in Florida sighted a cattle egret near his home. But he figured it must be an escaped bird so he did not report it. It was probably the first sighted record of the bird in this country. The important thing was that the 1952 birds arrived in large enough numbers to set up housekeeping.

The first nest turned up in a heron rookery at Lake Okeechobee on the fifth of May, 1953. It contained one egg. Three more nests were found later in the month, and although time didn't permit a more detailed search, the people who made the discovery estimated that there were at least a dozen pair.

From these beginnings, the species has spread out and today can be found nesting in Louisiana, South Carolina, New Jersey, New York, as well as Florida. Other states will surely join this list before long.

The cattle egret is often found in Europe and Asia, and it probably crossed the Atlantic on storm winds. It became established in British Guiana in 1877 and eventually spread to countries of South America and later to the United States.

Cattle egrets arrived in America on their own and are spreading rapidly throughout the land.▶

After its 1952 invasion of North America it spread widely and has now been recorded from much of eastern United States, the middle west and even in California. It is truly a bird on the march. The National Audubon Society's annual Christmas Bird Count first listed the bird in 1953. Only eight years later in 1961, the count numbered 3,120. Bird students in Texas recorded 10 pairs of cattle egrets in 1959; in 1966 the count was up to 20,000 pairs!

The United States already has a good number of heron and egret species. Does the cattle egret pose a threat to one of these? Will it prove stronger and dispossess some native American birds? Such are legitimate questions to ask whenever a new species adopts the American wildlife scene.

Fortunately the answer to both these questions appears to be "no." The cattle egret usually breeds somewhat later than the other species and uses smaller areas among the branches for making its nest. Furthermore, the cattle egret is primarily a bird of uplands and open fields rather than of the marshes. Therefore it does not generally compete with other birds for food or a place to live.

Cattle egrets feed largely upon insects. They move about in the grass catching such insects as they disturb. Throughout their range the birds have learned to associate with larger hooved animals, particularly the cattle for which they are named. Running about, beside, and under the big animals, the birds feed upon the insects stirred up as the mammals go about their daily business. The egret also picks insects from the cattle and frequently sits on the backs of resting cows. This bird has found a role in the ecology of the continent that has not been filled by any other.

The cattle egret is writing a unique new chapter in the bird history of this continent. It is the first species to arrive in this country from the Old World under its own power and to remain and multiply. Furthermore this vagabond has apparently found a place for itself in the ecology of the continent and does not pose a threat to some native species. We tentatively bid this gypsy welcome.

Cattle egrets catch insects stirred up by hooved animals.

SIX

That Man May Hunt

THE RINGNECKED PHEASANT

The perky little beagle was hot on the trail of a pheasant. He had followed the trail through the woodlot and out into the old field. Every little while the short-legged, little hound would make a great leap so it could see over the high grass. Now the bird was headed for the low marshy swale just beyond the stone wall. The dog was in hot pursuit.

It was a crisp October day and the trees were in their full autumn splendor of red and gold. The day was perfect for the boy on his first solo hunting expedition. It was also the dog's first real hunt with a gunner. The boy had run the young dog on pheasants all through the fall and had fired the gun around the dog so it would not be gun-shy. But this was the big day. Today they would put it all together.

The boy was nervous. Would he be able to hit a pheasant when it took to the air? The dog, he knew, would do its job of finding and chasing the bird. But could he hold up his end of the hunting partnership? He moved quickly down toward the stone wall to get in a better position for a shot. He guessed that once the bird cleared the wall it would take to the air.

From his new position, the boy could see the grass waving as the pheasant dodged and twisted downhill towards the wall. Seconds later the bird appeared at the wall and cleared it with a couple of wing-beats. Once on the other side it ducked under a clump of dogwood and crouched motionless. The boy moved into position for a clear shot. By now the dog was at the wall. For a brief moment it was confused by the break in the scent, then went over the wall.

The boy tensed in anticipation of the flushing bird. There was a sudden loud squawk! No thunder of wings followed. No bird rocketed skyward. Instead the dog appeared on the stone wall, tail wagging, the limp pheasant dangling from his mouth. The boy didn't know whether to cry or yell at the dog, or what. He had been cheated of his part in the hunting drama and the dog had performed a hunting dog sin. Men must do the killing, not the dog.

Thus went my first bird hunting expedition and one of my last. Our particular area of Connecticut just didn't provide enough habitat to produce anywhere near as many pheasants as there were pheasant hunters. So the State Fish and Game Department pen-raised thousands more and released them into the world at hunting season time. Those that were not shot in the first few weeks were fairly certain to fall prey to cats, foxes, weasels and other predators, or starve during the winter months. Losses of game farm birds are often 60 to 80 percent in the first six months following their release. Often the pen-raised birds were almost chicken tame. It was one of these semi-tame birds, that didn't have the sense to fly away, that was caught by my eager pup.

It was Benjamin Franklin's son-in-law, Richard Bache, who first brought some ringnecked pheasants to this country. He brought the birds to his New Jersey home from England around 1790. The birds just did not take hold probably because there was not enough open farm land in the area. Apparently no one tried again until 1880 after human activity had greatly changed the New Jersey landscape. The birds found things much more to their liking and were well established there by the early 1890's.

John C. Phillips brought a strain of nearly pure dark-necked English pheasants to his home area of North Beverly, Massachusetts in 1897–8. More than 100 years of farming throughout the New England region had created many good pheasant habitats and these birds soon spread into New Hampshire, Maine and Vermont.

The eastern pheasants, out of English stock, did reasonably well, but the first truly outstanding establishment of pheasants in America came in Oregon from Chinese stock. These birds were sent to the Williamette Valley of Oregon by Judge O. N. Denny who was Consul General of Shanghai. The first shipment he sent never arrived. Some people believe they were eaten by the ship's crew as part of a holiday celebration. Whatever the true facts, later shipments were made and did arrive in 1881. They were released and found conditions just to their liking. Within ten years the population was large enough for the state to open a hunting season on this bird.

Fish and Game Departments of many states were soon involved in bringing the ringnecked pheasant to their state. The various states set up large game farms to raise the birds, for it would take massive numbers of releases in order to have enough survive to establish wild populations. Initially, three races of pheasants were imported, the blacknecked from around the coast of the Black Sea, the Mongolian from Central Siberia, and the Chinese from eastern China. Most of our birds today are a blend of these races.

The ringnecked pheasant has prospered in the agricultural lands of the northern states. Here they have found a habitat essentially man-made where no native game bird has prospered. They do well in the fertile river valleys and bottom lands that are extensively farmed. The races of pheasants that have established themselves in the northern states have not taken hold in the southern states. Why this is so is not clearly known. In recent years, however, a pheasant race from northern Iran has been brought to the southern states and seems to be adjusting where its cousins have not. The pheasant may soon be as much a bird of the agricultural South as the agricultural North.

The best time to see ringnecked pheasants is early morning or late afternoon when they are doing most of their feeding. During midday they usually rest. In late winter and early spring the male birds establish a crowing territory for themselves. This may be anywhere from twenty to seventy acres in size. The males set up a number of crowing spots around the territory where they can "strut their stuff." At each of these spots they throw back

Male ringnecked pheasants have red face patches, dark green heads, and gold and russet bodies.▶

their heads and crow, then flap their wings vigorously. They then dash off to another of the crowing spots and repeat the performance. The main purpose of such activity seems to be letting other male pheasants know that this territory is occupied. Meanwhile the hen pheasants are selecting nesting grounds. When they are satisfied they mate with the male whose territory they have selected. One male may have anywhere from two to eight "wives," depending largely on the size and desirability of his territory.

Hen pheasants will nest in almost any vegetation that is high enough and thick enough to hide the nest and brooding mother. They like hayfields, fencerows, hedgerows, grainfields and orchards. The nest is a slight depression usually lined with grass. Most hens will lay around eleven eggs. Sometimes a nest will be found with twenty or more eggs, but more often than not this means that two hens are using the same nest.

The hens brood their eggs for 23 or 25 days. If the eggs are not destroyed by predators or early mowing or bad weather, about ninety percent of them will hatch. If her eggs are destroyed, the female will start nesting again in a new spot. Studies show that in good pheasant country 70 to 80 percent of the birds will nest successfully even though it may take several renestings.

Although great horned owls, Cooper's hawks, cats, foxes, weasels and skunks are all predators of pheasants along with man, some much smaller birds also provide a threat to the pheasant. Up to eight percent of pheasant eggs may be destroyed by crows, grackles and bluejays.

Probably the greatest natural threat to nesting success, however, is cold, wet weather during the first six to ten weeks of the nesting season. This is usually from May to September. Most birds will have finished incubating by the end of July, unless they have had to renest.

The adult ringnecked pheasant feeds largely on plant seeds, particularly farm grains and farm associated weed seeds. Of the latter, ragweed, foxtail grasses, wild sunflowers, and smartweeds are favorites. Such a diet built around plants of man-altered environments suggests why the pheasant has found a home here and presents little competition to native birds. In fact, it has undoubtedly helped other game birds by taking away hunting pressure that would otherwise have been directed at them.

With few important animal enemies other than man and being well adapted to agricultural lands intermixed with woodlots and wetlands, ringnecked pheasants have prospered in America now that the wilderness has been rolled back. As invited guests they retain their welcome. It is perhaps sad that they largely replace some birds such as the turkey, the prairie chicken, the heath hen, and the ruffed grouse; all of which decreased in numbers as the uninvited vagabond—man—increased his population.

SEVEN

A Bird of Royalty in America

MUTE SWAN

The canoe glided smoothly through the waters. My companion and I were enjoying the beauty of a lazy river on a lazy day. Purple clusters of pickerel weed blooms brightened the marsh along the shores. Yellow spatterdock blooms floated amid the green confetti of duckweed in the backwaters. Dragonflies darted back and forth in search of their insect prey.

As the canoe drifted around a bend, we saw that the river broadened out into a wide bay. It was very beautiful. The trees along the shore reflected in the still waters as did the brilliant blue sky and fleecy clouds. To put a finishing touch on this picture, a swan moved out from shore and began to glide gracefully across the bay in our direction.

Its neck was held in an S curve and its wings were raised in an arch over its back. As it got closer we could see its knobbed reddish-orange bill pointed downward. It was a mute swan. Although it was a beautiful sight, we were puzzled. We knew almost all the people along the river and none of them owned swans. We decided to move closer to the bird. It did not seem afraid as we paddled

A BIRD OF ROYALTY IN AMERICA / 75

carefully toward it. In fact it seemed to increase its speed in our direction.

We moved even closer enjoying the grace and dignity of this large white bird. Then suddenly it threw its head and neck back between its wing feathers. The whole body of the swan seemed to fluff up. We dug our paddles into the water to brake the canoe. We tried to turn the canoe. The swan paddled furiously in our direction with a loud hissing, thrust forth its neck and charged our craft, beating his wings as it came. The canoe turned sharply but not before the swan delivered a wrenching pinch with its bill on my paddle arm. We dug hard into the water with the paddles and the canoe shot ahead with the swan in fast pursuit. After a few yards it slowed down and again put its neck between its wings, puffed its feathers and sailed on. The swan turned and headed back toward shore.

In a few moments a second swan appeared followed by six fluffy grey young. The male turned and made another threat in our direction, then joined the family as they swam to the other end of the bay. What had started out as an idle and peaceful day had ended in a panicked exit for us and a painful arm for me. At least it was an experience not soon to be forgotten.

We saw the swan family often after that day. We took great pains, however, to leave plenty of open water between us. The sight of the cob, as the male bird is called, patrolling for enemies was always a handsome one. We had already learned the hard way that he would actively defend his family if he thought they were threatened. Angry cobs can be quite ferocious. There are records of mute swans in English parks attacking young children and dogs that approached the family too closely. Dogs have even been grabbed and dragged into the water and drowned. We treated this cob with great respect.

We learned that although the mute swan is the species found in our parks and on some large estates, this family was wild. There are now thousands of them from Rhode Island south to Maryland and Delaware. Before the mute swan was brought to America only the handsome whistling swan appeared along the east coast.

The mute swan is found wild in Denmark, other parts of northern Europe, and western Asia. It has long been domesticated because of its grace and beauty. In England it is considered a royal bird. No British citizen can own one except by special permission of the king or queen. Those who are given permission to keep swans are also given a "swan-mark." This is a letter or some device that is cut into the upper part of the beak. It establishes ownership in the same way as branding a cow. The custom of swan-marks goes back to 1482 and the reign of King Edward IV.

Even today, the swans along the Thames River in England belong to the Queen and to two city companies, the Dyers Company and the Winters Company. Every year a "swan-upping" is held. We would probably call it a round-up. It is done with great ceremony. Special uniforms are even worn by the Swan-Uppers who catch the birds. When they are caught, the swan-marks are freshened up on the adult birds and put on the cygnets, or young birds. The cygnets also have their wings fixed so they cannot fly.

In America, city parks were often patterned after European parks. They were not complete without the royal bird—the mute swan. Around the turn of the 20th century there were a number of people in the New York City area who had made great fortunes. They thought of themselves as being in much the same class as the great nobles of England. Many bought large estates and built great mansions, particularly out on Long Island or up the Hudson River in Dutchess County. Quite a few of these estates had large ponds or lakes or a quiet river flowing through the property. Swans on these waters put a finishing touch to these great estates.

Two hundred and sixteen swans were imported for such purposes in the spring of 1910. Three hundred and twenty-eight came in the spring of 1912. Many more have been imported since that time. However, in America the tradition of swan-upping does not exist. No one caught the cygnets and fixed their wings so they could not fly. These birds left the parks and estates and took up a free life on the Hudson and Long Island Sound.

Mute swans pair for life. During the mating season the young cobs fight mightily for a mate. It is a matter of large importance in their lives.

The females, or pens, do not lay eggs until their second year. In fact, they often do not lay eggs until they are three years old. During their first year of breeding they will probably lay only three to five eggs. In their second year of nesting they may lay as many as seven. By the time they are fully mature the number of eggs laid will vary from nine to 11.

This majestic swan is the same old male that attacked the author in his canoe.▶

The swan nest is a bulky thing built of a variety of water and marsh plants. It is always built very close to the water for swans dislike coming out on land. Incubation of the eggs takes about five weeks; longer if the weather is cold. The pen does all the incubating. The cob stays close by on the alert to drive off any intruders. Should something happen to the pen the cob will finish incubating and rear the young.

Although the bird is called mute, it is not truly silent. It may lack the great bugling voice of the trumpeter swan but it has sounds of its own. As a family sails along, the downy young use a number of soft whistling calls. If the family is separated the adults call with a sound like the barking of a little dog. And, of course, it hisses viciously on the attack.

The numbers of wild swans have been growing, but it takes some time for the population to grow since the birds take a relatively long time to mature. The young birds need two or more years before they lose their dingy grayish feathers and take on the full white adult plumage. It may be six years before the females are laying the maximum number of eggs. Of course, the birds live 30 to 40 years, so they produce many young over a lifetime.

At first, the swans were limited to the Hudson River and Long Island Sound. The birds from up-river, in Dutchess County, migrated south each year to warmer areas. As the numbers of swans increased, they spread out up and down the coast in both directions from New York.

Except for its ferocious attacks on people who carelessly approach the nest or young, the mute swan does not seem to be an additional problem to our wildlife. It will drive off ducks and geese that come too close to its nest. But fortunately the areas where it now lives are not high producers of waterfowl. It does not compete with the native whistling swan which nests farther north.

Like other successful animal immigrants, the mute swan has found an open ecological niche not filled by a native species. It is the only species of introduced waterfowl to have found a home here so far. It is certainly a very beautiful addition. To watch a pair of these birds sailing over the water, often with the downy young riding on the mother's back, is a joyful experience. It is truly a royal bird.

EIGHT
They Turned the Tide
STARLING AND HOUSE SPARROW

They were The American Acclimatization Society of New York and they had a dream. They planned to introduce to America all the birds mentioned in the writings of the great poet and playwright, William Shakespeare. They were gathered together on this day, March 16, 1890, to release 40 pairs of starlings.

As the birds dashed from their crates and circled up among the trees of Central Park, they were watched with great concern by members of the Society. Would the birds survive? That was the big question. Other clubs had tried and failed, but perhaps the New York club would be lucky.

There was great joy at the American Acclimatization Society when a few weeks later they received reports that a pair of the starlings were nesting under the eaves of the American Museum of Natural History. A close watch was kept on the nest. The eggs hatched and the young survived. This was a good omen. Just to be sure, the club brought in 400 more of the birds and released them in Central Park the following year.

No question about it, the New York birds thrived and multiplied.

Several pairs of starlings bred in 1891 with good success and only four years later the bird was common in the general area of New York City and Long Island. From this beginning the number of starlings has grown to countless millions, spreading across the nation to California and northward into Canada and Alaska, and as far south as southern Mexico. It has truly captured the continent.

The starling is an aggressive but pretty bird that is considered by some a welcome addition and by others a terrible pest. As common as the bird is, many people never look to see its changes throughout the year.

In spring this short-tailed blackbird is at its most beautiful with its bright yellow beak and purplish and greenish gloss to the feathers. In winter the birds are speckled with white. After the breeding season the sharp long beak of this bird turns from yellow to dark brown. During the winter it slowly turns to bright yellow again. During late summer you may see starling shaped birds that are grayish brown. These are the young birds. Later in the fall they will slowly change to an adult spotted plumage. Thus you may find birds with brown heads and dark spotted bodies.

Starlings are not great singers, but they are good imitators. Some of their own notes are rasping and harsh although they do have some musical whistles. They mimic, not only other birds, but barking dogs and mewing cats. Many a pretty young woman has taken a quick look for the maker of an admiring "wolf" whistle only to find a disinterested starling making the call.

The starling is very prolific. In this country it usually raises two broods of young each year and sometimes even three. The nest is built in any cavity high or low. They have nested only two or three feet from the ground in a hollow apple tree and more than 75 feet high in church steeples. It is not uncommon for them to use bird houses and drive out the birds for which the house was intended.

Starlings are often a plague to other hole nesting birds. They push out bluebirds, swallows, martins, house wrens and even occasionally flickers from holes and nest boxes. Often they destroy the eggs and young of these species, and conflict with native birds in another way. They eat the winter berries early in the season before they move south. This leaves little important food for the winter birds to eat.

Farmers and fruit growers also experience the destructive abilities of starlings, as they are particularly fond of cherries, strawberries and grapes. A flock can completely strip a cherry tree of its fruit in a few minutes. These birds will pick holes in apples, pears and peaches. Ripening tomatoes also attract them. As does green corn ears, sprouting corn seeds, young sprouting plants of peas, radishes, spinach and lettuce, and even planted seeds that have not yet sprouted.

On the other hand, their insect eating during the summer months is very beneficial. They are particularly valuable because they feed on introduced pests like the gypsy moth that many native birds ignore. The editor of this book told me of watching starlings in his backyard that were so gorged with gypsy moth caterpillars they had great difficulty getting airborne. Starlings will also feed

Besides eating fruit crops, the starling feeds on many kinds of harmful insects.

heavily on the grubs of Japanese beetles that are munching away at the grass roots in people's lawns. Several scientists have declared that the starling is the most effective enemy of the clover weevil in America. They are also known to eat many other serious insect pests.

The food needs of starlings take them to open grasslands, particularly lawns and pastures. They are easily recognized among other blackbirds such as the cowbird by their waddling gait and zigzag path. When they fly they flutter for a moment and then sail. Flutter and sail.

Although they fly out to the countryside during the day they usually return to the cities at night. The flocks come to favorite roosting places such as building ledges, bridges and sidewalk trees. After a day of feeding they have produced much solid waste. Their droppings mess up buildings and make it very unpleasant for people walking underneath the trees. Their constant squeaking, chattering, and grating calls as they jockey for a favorite spot add to the already excessive city noise.

If there were only a few of the birds it might be tolerable but their numbers have grown unbelievably. In 1948, famed bird watcher Roger Tory Peterson wrote the following:

"In recent years, as I sat at my desk on the sixth floor of Audubon House at 1000 Fifth Avenue, New York, on cold winter afternoons, it seemed incredible that the black blizzard of birds that swarmed to the ledges of the Metropolitan Museum across the street could be descended from a mere ten dozen birds released in that neighborhood hardly more than 50 years before. On several afternoons I put on my overcoat and stepped out onto the small balcony to watch the evening flight, the vanguard straggled in after four o'clock. By 4:30 large flocks began to sweep in. During the next half hour, flock after flock, 300 or 400 at a time, poured through the gap in the buildings above me. Most of them came from far out on Long Island, where commuters see the city bound flocks 20 miles or more from Manhattan. Other flocks come down Fifth Avenue from the direction of the Bronx. A few arrived from the west. It was like watching a three ring circus. Yet there was an orderliness about it all. Each flock handled itself with military precision, like a squadron of fast aircraft. When one bird turned they all turned. When a flock passed in front of another flying in a different direction the sky was crosshatched with a moving pattern of birds. I tried counting but the starlings pitched on to ledges so fast I had to guess at the number of each incoming flock. At times the sky was a swirling storm of wings and even the broadest estimate was pure guessing. The first evening when I totalled my figures the count came to 24,000. A week later, when I tried it again, I got 36,000. The din from the thousands that shoved and jostled on the ledges could be heard far into the night. They never seemed to sleep. When I passed by on a Fifth Avenue bus at midnight, they were still squealing."

Peterson says later in his book that this was a small roost. He estimates the roost along Pennsylvania Avenue in Washington to be greater than 100,000 birds. Such roosts of many thousand starlings are common to cities across the nation. They also roost in groves of trees and large marshes. In such places they are often accompanied by great numbers of grackles, redwings, cowbirds and other blackbirds. Many such autumn roosts have more than a million birds in them. The birds coming in at sunset resemble great swirling rivers in the sky. Such rivers are many miles long and thick enough to block out the sun at times. We have seen nothing like these great flocks since the passenger pigeon vanished from our skies forever.

All manner of devices have been used to discourage these great roosts in our cities. Rattles, balloons, fireworks, whistles, stuffed owls, firehoses and blank shells have been used to drive them off. Recordings of their distress calls have been made and played back to them. Bare electric wires have been placed on roosting sites as has a series of sticky chemical repellents. All have discouraged the birds very briefly but they are highly adaptable and soon return or move to nearby areas. The starling remains the winner. In Englewood, New Jersey, city officials cut down the trees where the birds could roost. But who won?

Despite debates in city halls, state legislatures and the halls of Congress concerning what to do about this bird, the starling goes right on multiplying. Each year it causes millions of dollars in crop losses, but the starlings also consume billions of insect pests each year as well. Is this bird a blessing or a curse? It is hard to say for sure. Certainly we could do with far fewer of them. They are

A roost of starlings may be seen on city trees and building ledges at sunset when they return from feeding in the countryside.

unprotected by law and are a good sport for the hunter for they make a difficult target. Very few people hunt them, however. Perhaps they would if they knew what good eating they are wrapped in bacon, roasted, and served on toast.

THE HOUSE SPARROW

It was the success of the starling in America and of the house sparrow that awakened officials to the potential dangers of deliberately importing exotic animals. Although most species did not survive, those that did could be too successful. The house sparrow, once it became established, showed a massive population explosion just like the starling, but after the invention of the automobile its numbers have decreased.

As with the starling, the first groups of English, or house, sparrows did not take hold in America. Eight pairs were brought from England to Brooklyn in 1850. They were kept caged through the winter and released in the spring. They did not survive. It was decided to try again. Nicholas Pike got a second shipment from England and released them in Greenwood Cemetery. This attempt was successful. The birds did well and multiplied. A bird of city and farmyard, man-shaped environments, the house sparrow found homes all across America within a few years. In fact this bird, which is actually neither English nor a sparrow, is a weaver finch. It is native to many parts of the world and does well anywhere outside the tropics. It is probably known by more people than any other bird.

Within forty years after they arrived, the house sparrows were an important part of the American scene. They do not generally move very far from the places where they were born. Two to six miles is about as far as they usually go, although an occasional individual will travel more than 200 miles. How did they spread across the country so rapidly then? Some, of course, were imported but others traveled the railroads like hobos. Empty railroad cars often had scattered grain in them from their cargoes. House sparrows entered them to feed. Often they would be closed into the empty cars by accident. When the doors opened, car and sparrows would be in a different town miles away. Here the birds would soon establish themselves as comfortably as before.

The weaver bird family, to which the house sparrow belongs, is known for its beautifully woven and complex nests. The house sparrow, however, is hardly the pride of the family when it comes to nest building. It builds a clumsy ball of dead grass or straw which it lines with softer materials: hair, string, fine grass and feathers. The nest is about the size of a football and has a hole in the side. Actually, tree nests are rather uncommon today; most house sparrows prefer to nest in holes. But even in holes they build the same kind of domed nest.

They have become very fond of bird houses as nest sites. This has made them unpopular with people who wish to entice bluebirds, tree swallows, and purple martins to their homes. House sparrows nest earlier than those other birds so they have taken over the available boxes by the time the others arrive north. They defend their nests savagely. The others have no choice but to look elsewhere. Unfortunately, with the fast disappearance of dead trees with holes from the countryside, there is often nowhere else to turn.

House sparrows also nest under eaves, in gutters, and other suitable sites on man-made buildings. The females lay three to seven white eggs splashed with brown or gray. The youngsters are noisy, and both young and parents are often messy by human standards. These birds are not only aggressive to others, but are very scrappy toward members of their own species. Females take more than one mate and the males battle heartily over the females. Since breeding goes on almost throughout the year, there is usually some scrapping going on among a colony of these birds. The cheeping and chirruping calls that accompany all the activity are not very musical and do little to make the bird more attractive to humans.

Although the birds were brought here originally to control an insect, the Linden Looper, it was a foolish idea. The birds do eat insects but they are primarily grain eaters. Indeed, the house sparrow prefers land with a mixture of trees and grass. Man generally seems to prefer the same kind of habitat so the house sparrow has naturally become associated with him. Man is a maker of artificial grasslands—his grainfields, parks and lawns with trees and shrubs near his home. This all adds up to an open invitation to these birds and they eagerly accept.

In the early years of the house sparrows' move to America, the horse and buggy were the main forms of local transportation. There were many stables where the birds could get grain and there was always some undigested grain in the horse droppings in the street. The abundance of such food made life very easy for them. They multiplied tremendously and probably reached their peak numbers around the turn of the century.

After that time a number of factors have combined to reduce the populations to about one-quarter of the peak years'. When the gasoline engine in cars and trucks replaced the horse, the large grain supply disappeared. Although the sparrows learned to eat dead insects from the radiator grilles on the early cars, that was no substitute for

The house, or English, sparrow frequently nests in bird houses or under the eves of buildings.

grain. The starling also gave the sparrow competition for some nest sites.

American predators, like the sparrow hawk, learned to hunt the sparrow in the countryside. Better farm practices reduced the amount of waste grain available to the birds and the tractor replaced the horse on the farm as the car had replaced it in the city. There is no doubt the house sparrow is here to stay but it will not see its golden age of the 1900's again.

As proof of its permanence we have evidence that a number of varieties are developing in a pattern similar to that of native wildlife. The house sparrows of the northern states are larger, darker birds than the house sparrows of the arid Southwest. It means that house sparrows in the United States have separated into subspecies. The separation was achieved through evolution, and all within 110 years, much of it since 1900! Yet these changes fit within the rules that scientists have discovered among nonmigratory creatures. For instance, a northern population of birds will be relatively large, since a larger body holds heat against the winter cold; whereas in hot, dry regions the body would be small, getting rid of heat faster.

The house sparrow has filled an ecological niche that was made by man and was vacant. No native species had stepped into it. Today, the sparrow is spread across the nation and is slowly fitting itself more exactly to local conditions.

The starling and the house sparrow successes resulted not only in awakening leaders to the dangers of importing new species but to laws to control importation. In 1900, Congress passed the Lacey Act regulating the importation of exotic species. It says, in part:

> "Sec. 241. The importation into the United States, or any territory or District thereof, of the mongoose, the so-called "flying foxes," or fruit bat, the English sparrow, the starling, and such other birds and animals as the Secretary of Agriculture may from time to time declare to be injurious to the interests of agriculture or horticulture, is hereby prohibited; and all such animals and birds shall, upon arrival at any port of the United States, be destroyed or returned at the expense of the owner. No person shall import into the United States or

into any territory or District thereof any foreign wild animal or bird, except under special permit from the Secretary of Agriculture."

This act has been a help but it has not prevented permits being granted without adequate research or the release of small animal pets, particularly reptiles and tropical fish, in states like Florida and Hawaii.

ALIEN

FISHES

Fishes are the oldest of vertebrate animals and have been around this planet a long time. Probably the greatest number of fishes live in the vast oceans but a large host of species live in the freshwaters of the world. For most fishes land is a barrier, and so a new species is usually confined to the watershed in which it evolved. For instance, many fishes that developed in the Mississippi River drainage are found nowhere else in America unless they have been carried to other river systems by man and there released.

Such action has not been uncommon, particularly with sport fishes such as trout. Rainbows, brook trout, Dolly Vardens and others have been "stocked" in many waters across America. Indeed even in Europe, North Africa and South America, ardent fishermen cannot bear to be any further removed from their favorite fish than is absolutely necessary.

Pan fishes, such as many varieties of sunfishes, bluegills, and crappies, have likewise been man-moved around the country, usually deliberately. Some of our common minnows have hitchhiked around this country although with less of a direct invitation. Most arrived as bait fish to be used in catching their more glamorous cousins. Some escaped the hook. More were just dumped into the water at the end of fishing expeditions. Since these "bait fish" often eat the eggs of game fish and compete with their young for food, they are declared by fish managers to be "trash fish." Such managers try to get these intruders out of pond and stream and many states have severe limitations on fishing with "live bait."

All of the fish in the following section, however, are newcomers to the waters of North America. They were brought here originally for food, sport or the aquarium hobby. Some have proved something of a blessing; others are more of a curse.

The raising of quantities of tropical fishes for sale to the tropical fish hobbyist has presented very special problems to the state of Florida. The subtropical climate of the area proves quite suitable to many species of fish, and so there are many "fish ranches" in the state. Periodic hurricanes may cause even the most carefully designed pools to flood and spread their occupants into the wild. It would take too long to list the many kinds that have established themselves in one drainage system or another throughout the state. Added to these escapees are the deliberate additions of species such as guppies, mollies and others to ponds for the control of mosquitoes. Probably no other state in the nation sports the variety of non-native fishes that Florida does.

NINE

Introducing the Biggest Minnow

THE CARP

Carp are the largest of the minnows. Now if this sounds a bit strange since most people think of minnows as any little fish, I should perhaps point out that true minnows actually are a member of a particular family of fish. Scientists call the family *Cyprinidae*. In North America alone, there are 193 species in the family. It is true that most minnows are rather small but the carp, known in the scientific world as *Cyprinus carpio*, has been known to grow four feet long and weigh 50 pounds.

As a teenager, I had fun fishing with a bow and arrow. And it was not uncommon to catch five pounders that ran more than 30 inches long. While paddling the canoe into the shallow, muddy backwaters of the Housatonic, we were alert for the V-shaped ripple of a carp's back as it cruised along searching for food such as aquatic plants, insects, crayfish, snails, and small fish. One of us would paddle while the other stood with drawn bow on unsteady legs as the distance between canoe and carp closed. Twang! The arrow would reach out toward the fleeing fish trailing its hindering line. Usually the arrow zopped into the water and bobbed up uselessly to the surface seconds later. Occasionally we actually connected and then the fun began, for a reel on a bow is not quite the same as one on a fishing pole. Now a carp is not the fighter that a bass or tarpon is, but some of them took us for rides in the canoe that made us feel a bit like whalers being taken in tow by a big whale for what they called a "Nantucket sleighride."

In some parts of the country carp are pursued as a game fish fairly regularly. Ten and 15 pound carp are not uncommon and in the big lakes of the midwest 30 pounders are not unheard of. What it lacks in fight, the carp makes up for in size. On July 10, 1952, Mr. F. J. Ledwein of Clearwater Lake, Minnesota, came home with the record carp taken on rod and reel in U.S. waters. It was 42 inches long and weighed 55 pounds five ounces. Big as that is, heavier ones have been caught in other parts of the world; for example, an 83 pound specimen was hooked in South Africa.

Carp have long been raised for food in Asia and Europe. They were brought here originally to establish commercial carp fishing. The first attempts to bring in carp began in the 1830's but no permanent population was established. Importations from Germany to California in the 1870's were more successful, however. Today the fish can be found across the entire continental United States, southern Canada and Mexico. It does provide commercial fishing as hoped and in the midwestern United States it is the major commercial species. About ten million pounds of carp are marketed in the United States each year, and around a million pounds a year in Canada.

The carp looks very much like a large version of its "living room" cousin, the common goldfish. Among normal carp there are two distinctly different types that appear now and then. One lacks scales completely and is called "leather carp;" the other has patches of the large thick scales on the back and sides and has been named "mirror carp." The offspring of these are usually normally scaled fish. It is estimated that today about two percent of our wild carp are one of these types.

Around the mouth of the carp are two pair of barbels, little fleshy points that are full of nerve endings. These help the fish find food in muddy waters where eyesight is greatly hindered. Muddy waters and carp go hand in hand. If the water is not muddy when the carp arrives it will be before the fish is there very long. It is largely this habit which has made the carp very much of a mixed blessing as an addition to North American waters.

The silt the carp stirs up may smother the eggs of other fish and its uprooting of plants destroys hiding places for young fish. It also eats the eggs and young of other species, and is suspected of spreading parasites and disease. Thus the carp is hated by many anglers and conservationists because of its competition with game species for food and space and its roiling of waters.

Other people argue that the carp generally lives in waters too muddy for other species. It also is fairly tolerant of polluted waters and will feed on sewage and the small creatures such as bloodworms (midge larvae) and rat-tail maggots that also find a home in domestic waters. These people point out that the carp offers people the joys of

The carp is a relative of the common goldfish.▶

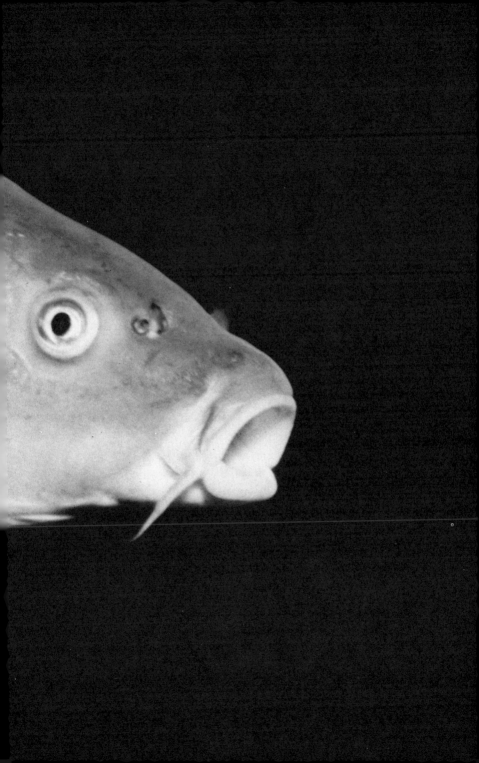

fishing where no other fish are to be found. This may be true, undoubtedly it is, but somehow it seems a sad commentary on what has happened to American streams, lakes and ponds.

A large female carp will lay over a million eggs in a season. She selects as her nursery a shallow weedy bay or tributary of a lake or the backwaters of a river sometime from May to the middle of July. Coursing back and forth with the male alongside, she scatters her eggs at random over the plants and muck. The male spews forth his milt, or sperm, at the same time. Both parents abandon the eggs as soon as their mating is over.

The eggs are slightly adhesive and stick to the plants on the bottom. Those that are fertilized by the milt hatch from six days to two weeks later. They encounter many hazards that face all young fish—predators, disease, temperature changes and pollutants.

Those that make it to adulthood may provide food or sport for people—particularly city people who live along the polluted waterways that are now in the majority in this country. The carp may soon be the only wild fish they know. Future generations, indeed many of today's, will go down to the stream armed with dough balls, raw potato cubes or earthworms in search of the wily carp. Unfortunately they will find many signs that warn them: *Don't drink the water. Don't eat the fish.*

Because the carp, more than most other fish, is tolerant of pollution, their presence in a body of water is an indication that all is not well. Without meaning disrespect to the "greatest minnow," I hope to see it decline in numbers due to the cleanup of America's waterways.

TEN
The Fish That Runs Away
WALKING CATFISH

The biologists of the Florida Game and Fresh Water Fish Commission figured they had a new problem on their hands. They discovered that the walking catfish (*Clarias batrachus*) of Southeast Asia was making a successful invasion of Florida. Any introduction of an alien creature is potentially a problem.

The biologists set out quickly to contain the invasion before it got out of hand. They used a tried and true tactic of preventing the spread of an unwanted fish species. They located the ponds where the fish were known to occur and poisoned the water with rotenone. This is usually very effective in killing most fish including Clarias, the walking catfish. But it isn't called the walking catfish for nothing. This species avoided the deadly chemical by crawling out of the treated ponds and slithering away to more hospitable waters.

Clarias have an excellent appetite and an aggressive disposition to match. Adults have been observed feeding on freshwater shrimp, crayfish, tadpoles and snails. This diet would not affect other native fishes too badly except that the walking catfish would be one more competitor. However, walking catfishes by no means stop at these food items. They also consume quantities of panfishes such as bluegills, warmouths, dollar sunfishes, and native minnows. A relative, the native catfish, is also included in the diet. And one 13 inch Clarias was observed killing and eating a seven and three-quarter inch bullhead.

Young Clarias from five to seven inches in size have been studied to find out their food preferences. Fish made up the bulk of their diet, but they also ate a great deal of insects and insect relatives. Florida waters abound with such fare. Clarias should find plenty of food to build their numbers and foster their spread to other areas.

This new addition to the Florida fish fauna was first brought to the attention of fishery biologists on March 15, 1967. On that day a strange fish was caught and brought to the staff of the Loxahatchee Refuge in Palm Beach County for identification. The U.S. Fish and Wildlife Service people at Loxahatchee turned the oddity over to the fishery biologists of the Florida·Game and Fresh Water Fish Commission.

They conducted a field investigation in the area, but found no more specimens of walking catfish.

Seven months later, fishermen claimed to have seen another Clarias in the same area but, since it was only a rumor, no field investigation followed. Then on May 25, 1968, a night watchman near Boca Raton went to investigate the yapping of his dog. Imagine his startled surprise to find the dog confronting a ten inch pale pink catfish crawling along the bare ground. The fish was caught and

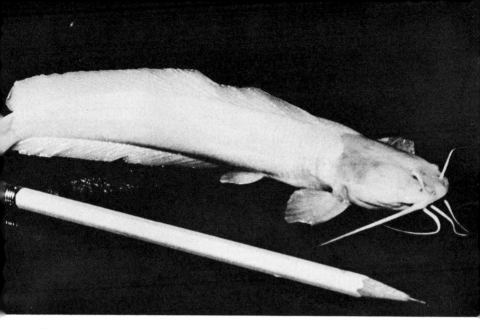

The pale pink walking catfish can wiggle over land for long distances.

a field investigation was launched. In less than a month six more adult specimens had been located in the area.

Where had these ghostly creatures come from? Their native home is Thailand. There they are widely distributed and are even raised in ponds as a human food source. However, they did not get invited to this country as food fish but as a curiosity for the huge tropical fish hobby that absorbs millions of Americans.

The vast majority of walking catfish are dark colored, varying from pale slate gray to mottled browns and blacks. Occasionally white (actually a pale pink) ones appear in the population. Fish fanciers preferred the ghostly ones and developed a strain of them for sale in this country. Florida, with its near tropical waters, allows fish dealers to raise fish in outside pools throughout the year. Most are responsible people and their pools are screened to prevent their captives from escaping into wild waters. Of course most fish merely swim—they don't climb out and walk away like Clarias. These fish have become the dominant ones in many ponds. In one small area of water 12 feet by 15 feet, only slightly larger than an average living room, two hauls brought in 65 Clarias!

The Florida authorities waged a valiant battle to stop the spread of this alien creature. In fact they had even planned to eradicate it. But when a fish refuses to swim around and die gracefully from poisons administered by man—man may have to face defeat. And so in November of 1968, the biologists admitted temporary defeat—the fish was already too widespread to eradicate, and available controls didn't affect the creature.

Just what kind of beast is this that has successfully infiltrated the state of Florida? Its homeland is actually most of eastern India and Southeast Asia where it regularly grows to around 22 inches. However, up to 1969, in the United States the largest recorded Clarias had been 18 inches.

The fish has a long eel-like body with a large flattened head that is preceded by eight long barbels. These it uses to feel its way around the bottom of the murky ponds where it dwells.

When ready to lay eggs, Clarias dig horizontal burrows into the bank some eight to 24 inches below the surface of a pond. Here 2,000 to 15,000 yellowish brown eggs are deposited. The eggs do best in water between 77° F and

90° F and at these temperatures they hatch in about 20 hours. Located in the large head are a series of sacs surrounded by blood vessels. These sacs are like the endless caves called *labyrinthines* and so scientists call the collection of the sacs the labyrinthines organ. This is the organ which allows the fish to gulp air and use the oxygen it contains while wriggling overland from one pond to another. In order to work properly the organ must be kept moist, and so too must the skin of the fish. Therefore Clarias goes abroad mainly at night when there is no sun to dry it out or during rainy periods. The fish can live out of water for many hours and it can survive in waters with a very low oxygen content. It can also live in brackish water.

In Thailand, Clarias eat worms, shrimps, insects and decaying animal material. It seems to prefer the latter. When pond-reared for food, Clarias are fed rice mixed with rice bran, fish meal or chopped vegetables. Perhaps in this country the walking catfish will provide a new food source that will give us both much needed protein and will enable us to control it as a potential pest.

How will this strange and interesting fish fare in America over the next decade or so? Will native predators and parasites "discover" it and develop a set of checks and balances for Clarias? Will it spread so fast that it wipes out several native species? No one knows for sure at this writing. Only time will tell.

ELEVEN

The Brook Trout Gains a Rival

BROWN TROUT

"What do you suppose those boys down at the State Fish Hatchery have in mind, George, bringin' in those cussed brown trout?"

"I don't rightly know, Ralph, but I sure figure that their plans aren't going to be any good for our ole native brookies."

"I've heard tell it's all that Fred Mather fella's doin'. He got quite a kick out fishin' for the brown trout over in Europe—you know, messin' around with them dry flies like old Isaac Walton."

"Well they can think what they want, Mather or no Mather. Releasin' them brown trout around here is only goin' to mean trouble. They may be bigger than brookies, but just because of that, they're going to eat up the young brookies. Mark my words, them cannibals will push the brookies right out of their streams. And then what kind of trout fishin' we goin' to be left with—them browns ain't got the fight nor the flavor of our little native brookie gems. Besides, who wants to stand around flickin' flies and tanglin' lines?"

"It's about what you can expect when some of those wealthy fellas get a hold of somebody in government. They forget about us little fellas. Besides, any American fish has got to be better'n anything they can bring over from Europe."

Conversations like this were certainly commonplace among fishermen shortly after brown trout were released into New York State waters in the late 1800's. The average fisherman believed the browns were a coarse fish, lacking in sport value, and cannibalistic to boot, sure to destroy the beloved native brook trout.

Fred Mather, a New York State conservationist, definitely had thought otherwise. The idle gossip was correct; Mather had enjoyed stalking the brown trout in Europe using the dry fly technique. He felt the brown would be a good addition to the trout family in America. Mather heard about a German named Von Behr who could get eggs for export. A deal was made.

The brown trout found the streams in America to its liking. It thrived and multiplied, and its critics were right. It did sometimes move into brookie waters and eat the smaller fish and it bred in some of the same areas brookies had used. The brown and the brookie don't seem to be able to peaceably coexist.

However, the critics greatly underestimated the qualities of the brown trout itself. The brown prospered in waters far too warm for the brookie. Brookies must have a high oxygen content in the water to survive and that means cold water. Furthermore, brown trout tolerate mild pollution that no brook trout could survive. Thus the brown trout took over eastern waters that had never known the brook trout. More waters could now offer trout fishing.

The brown was not the sluggish fighter it had been pictured to be. It fought well and sometimes it would leap from the water in its fight. The brook trout almost never did this. The brown was a better surface feeder and readily took flies and grew much larger than the honorable brookie. Of course, it was not blessed with the same jewel-like beauty but then again it was hardly ugly.

As its qualities became known, the brown trout was invited to waters across the country—invading the home of other trout as well as the brookie. Today the brown trout has a wider range in North America than any other member of the salmon family except the rainbow trout.

Today most brown trout are raised in hatcheries. However, there are wild populations in many streams. Wild brown trout show up less often in the fisherman's creel because they are more wily than their hatchery-reared cousins. They are also more beautiful. Hatchery-reared specimens tend to be dull and pale by comparison. In fact, a fisherman catching his first wild brown trout may think he has some slightly different variety.

The wild brown trout move up streams to their spawning grounds sometime between October and February depending upon where in the country they are living.

However, they do not usually get up into the very tiny rills and brooklets that the brook trout seek. Following a courtship ritual similar to other trout, the female scoops out a depression in the gravelly stream bottom using her

Brown trout are olive brown to bronze green on the back with bright yellow on the belly. Black, brown and red spots usually appear on the sides.

fins and body. Actually a large female may prepare three or four such nests over an area of twenty or thirty feet. When she is done she entices the male to the nest area. Then she begins laying her eggs. The male swims beside her and emits a cloud of milt to fertilize them. Their breeding chores over, the adults leave the area.

The female will have laid about 6,000 large non-adhesive eggs before her work is done and if the stream temperature is around 50°F, the eggs will hatch in about 50 days. In colder water it will take a bit longer. The young trout feed heavily on stream insects and crustaceans. As they grow larger, other fish increasingly become part of their diet.

Wild brown trout young face many dangers from enemies such as fish-eating birds and mammals, insects, disease, and pollution. They also have to compete with each other for food. Hatchery-reared trout have it much easier. At the hatchery the eggs and milt are carefully removed from the parent fish and mixed together to insure that the highest number of eggs possible get fertilized. The eggs are cared for in water treated to prevent any diseases that could affect the eggs. And the temperature is maintained at an optimal level. The hatchlings are carefully guarded from predators and disease in specially designed tanks and are given food enough for all. Far more eggs from each female progress all the way to adult fish in a hatchery than are ever the case in the wild. However, when these hatchery fish are released into a stream, their life is almost over. Those that don't end up in a fisherman's creel within a few weeks will usually die before long at the hands of predators, pollution, disease, or starvation. Only a handful make the full transition from hatchery living to a wild free state.

Although originally considered to be a nuisance, the brown trout has done much to improve trout fishing in America. There is no doubt that it has contributed to the decline of the native brook trout in many areas, but it certainly has not been anywhere near as much a factor as the warming of streams due to forest clearing and industry, and the polluting of its waters by many materials. Its greatest contribution has been that it has taken to waters that no brook trout ever called home.

ALIEN

INSECTS

There is something deep and mysterious about insects. They have known this planet for so much longer than man and his mammal relatives. It is not unlikely that they were roaming Earth probably 200 million years before our species arrived.

In addition, no other group of living things has so many different species. We don't even really know how many there are. Some estimate that there are between 2½ to 10 million individual insect species. As for the total number of insects, we cannot even guess. The actual count would be so large that we have no words in our language to describe it. To give you some idea of this figure try to imagine that on one acre of cultivated field, scientists have counted from one to ten million insects!

Not only that, but these creatures have fantastic reproductive rates. Fortunately, many have enough enemies to keep their multiplying numbers down. For example, it has been calculated that the offspring of one pair of houseflies in one summer would be 191,000,000,000,000,000,000 if all the eggs hatched and all the young survived.

From time to time, however, conditions are just right and a species reproduces faster than its enemies can control it. This over-population may exist for a year or two but usually the enemies catch up and bring the total back down. If this doesn't happen the numbers may very well grow beyond belief. Should the species feed on materials of value to man it will be called a pest. Then man must learn how to control the increasing population. He may even seek out the original home of the insect, round up some of its natural enemies, and bring them to the new area to feast on the troublemaker.

Among the roughly 83,000 insects in North America, there are quite a few that we consider *serious* pests because they interfere with human activities. Most of the species, however, do not affect man directly and there are a good many that are very useful. The latter includes the pollinators such as bees and flies, the weed eaters, and the eaters of harmful species.

The insects most harmful to human activities in North America are usually those native to some other part of the world. Most of them have arrived as stowaways aboard ships or planes. They have come on plants that have been imported, or on other products man uses for food, clothing or lumber. In this country, free of their usual enemies, and with abundant food, the numbers explode. A species that is only a minor nuisance in its native land may become a major economic pest here.

The chapters which follow look at several alien insects and their progress in this land in great detail. But there are many many others whose stories could also be told.

The boll weevil, about which songs have been sung and over which countless tears have been shed, pushed into the United States about 1892. By 1922 it claimed more than 85 percent of our Cotton Belt as home. India was originally home to the pink bollworm. It spread to other cotton raising regions of the world and laid claim to the United States beginning in 1917.

Cotton was hardly the only crop to be visited by alien insects.

Each winter the grapefruit and orange groves of the Rio Grande Valley in Texas are invaded from Mexico by the Mexican fruit fly. Its main way of getting about is on shipments of fruit from Mexico to the United States.

In Japan a beautiful bronze and green beetle exists that is not much of a pest. It arrived in New Jersey in 1916 probably on some plants imported from Japan. In this country the Japanese beetle found a paradise with many food plants, over 275 different kinds, and few enemies. Before long it was doing over ten million dollars a year damage to a variety of human interests. It ate farm and orchard crops; it made lacework out of prize flowers in backyard gardens. Its larvae ate the roots of grass on golf courses and lawns.

Studies were made in its native home. And with the aid of an imported fungus disease, chemical poisons, starlings and house sparrows, the numbers have been greatly reduced in recent years.

Corn in America faces considerable damage each year from the European corn borer which was first observed in market gardens around Boston. A bit of detective work by

This leafy lacework is evidence of Japanese beetles' hearty appetites.

Related to the sacred Egyptian scarab beetle, the Japanese beetle is a beautiful green and bronze. Nonetheless, man considers it a pest.

the staff of the Massachusetts Agricultural Experiment Station revealed that they had slipped into this country a few years earlier in a shipment of broom corn. This broom corn had come from either Italy or Hungary for use in a Medford, Massachusetts broom factory. By 1952 it had spread in 37 states which make up the bulk of our corn growing region. This insect causes several hundred million bushels worth of damage each year.

Before the 1920's, American towns were noted for their elm lined streets. Today, many, many towns are without any elms. Those that do remain are seriously threatened. The European elm bark beetle arrived in this country sometime in the 1890's in shipments of elm lumber. Since they ate only dead wood they were not considered pests. However, late in the 1920's the Dutch Elm Disease arrived here. It was a fungus disease. Its spores were carried from tree to tree by the beetles. Although the beetles were harmless, the fungus was deadly to elms. As carriers of the fungus the beetles suddenly became serious pests.

Not all of the introduced insects are harmful however. Quite the contrary. Honeybees are exceedingly important to pollination of our fruits, vegetables and flowers. Disagreement exists over whether there were honeybees in America before the white man came. Many suggest that the Indians learned to make maple syrup because they did not have access to honey as a sweetener. Others say, yes, America did have honeybees. The facts are, however, that most of the bees here today are varieties from European stock. For all practical purposes the honeybee is an alien species.

A great many parasitic insects have been brought here to fight some of the pests we have mentioned earlier, and predatory insects such as a species of ladybird beetles and the fascinating praying mantis have also been brought in for biological control.

Actually, America had one species of praying mantis living in the South but it was not abundant. Asiatic species and a European species were brought to the United States and survive much farther north. The mantis probably first arrived in this country accidentally but has also been deliberately imported in later years. The mantis eats whatever insects come its way.

The variety of introduced insects has really only been touched upon. The subject is worthy of a whole book by itself. The stories that follow are typical of the problems and possibilities that are becoming realities.

TWELVE

The Great Caterpillar War

THE GYPSY MOTH

"The street was black with caterpillars—so thick on the trees they were stuck together like cold macaroni. The huge hairy caterpillars were constantly dropping upon people on the sidewalks. The foliage was stripped from all the trees, and little was spared but the horsechestnut and grass in the fields, though even these were eaten to some extent."

Although it sounds like a quote from the script of a third rate horror movie it is actually a description of a scene in Medford, Massachusetts in 1880. The villain of the piece is the gypsy moth. This insect was deliberately brought to this country in 1868 by a French scientist, Professor Leopold Trouvelot.

Professor Trouvelot imported egg clusters to hatch and grow in experiments he was conducting. One of his hopes was to develop a hardy race of silk-producing insects. The common silkworm did not do well in cooler climates. The gypsy moth did. If he could cross breed the two insects, the resulting hybrid might be hardy as the gypsy moth and as silk productive as the silkworm. Had he been successful in his efforts, the Professor might have become a wealthy man. For in the days before artificial fibers like nylon and rayon, silk was one of the most expensive natural fibers.

To raise his caterpillars Professor Trouvelot placed them on branches of shrubs in his yard and tightly enclosed the branch in a bag of fine netting. When his precious insects had eaten all there was on the branch he would transfer them to another branch and again tie the net bag over them. This bag served two purposes; it kept the insects from escaping and predators from gobbling them up.

One morning the Professor awoke to find a fierce storm with gale winds playing havoc throughout the area. When the winds died down he went outside to inspect his insects. He was dismayed to discover many of his rearing nets hanging in rags and the insects widely scattered.

Trouvelot realized that this might be a more far reaching catastrophe than just the ruination of his current silk project. As a European, he knew the gypsy moth in its native land. He recognized that even under the control of its natural enemies the gypsy moth occasionally had population buildups that ended in serious defoliation of trees. What would it do when its natural enemies were absent?

The question was a significant one and one answer was that the moth would expand in numbers rapidly and become a serious pest consuming great quantities of tree leaves. Recognizing his responsibility for unleashing a potential pest onto the American continent, Professor Trouvelot promptly notified the public through the professional insect magazines of that time.

Unfortunately, no one seems to have heeded the warning and the insects ate their way through caterpillarhood, formed pupae and emerged as breeding adults. They laid their hairy, buff-colored clusters of eggs, each containing around four or five hundred eggs. The eggs hatched and the caterpillars munched their way to adulthood almost undisturbed by the native wildlife of the area. The gypsy moth multiplied very rapidly and by the late 1880's damage from the insect had become very noticeable in certain areas of Medford and nearby Malden. By 1889 the caterpillars had increased to such large numbers that they were completely stripping the trees in parts of southern Medford. With their food gone caterpillars swarmed out in all directions in search of food.

People in the neighborhoods were distressed. They didn't know what this pest was, for they had not read the scientific journals where Trouvelot had published his warnings. Specimens were collected and taken to the Hatch Experiment Station at Amherst, Massachusetts, where they were identified.

It quickly became obvious that the gypsy moth problem was too great to be solved by the local residents alone. They petitioned for legislation to exterminate the pest early in 1890. The state and local authorities took action. Governor Brackett presented the problem to the legislature which responded quickly. The governor signed the first act authorizing work against the moth on March 14, 1890. The act provided for $25,000 to do the job.

A three man commission was set up to do the work of fighting the insect. The commission soon discovered that the gypsy moth had already spread over a much greater area than had been supposed. This fact was brought to the legislature's attention and $25,000 more was appropriated on June 3, 1890. The war against the gypsy moth began in earnest.

The work was eventually transferred from the commission to the State Board of Agriculture. Crews were put to work to destroy the threatening insects. Egg clusters were painted with a combination of creosote and coal-tar pitch; trees were ringed with burlap and caterpillars seeking shelter from daytime heat gathered here where they could be killed; favored food plants were also removed in some areas. In 1893, the arsenal of weapons was increased by the use of lead arsenate. It was applied by use of hydraulic spraying machines from mid-May to mid-June when trees were well leafed out. This spray was toxic to animals, so sprayed areas had to be fenced to prevent grazing animals from being poisoned. However, the work was successful and the Board had apparently done its work. By 1899, the gypsy moth had been eliminated from many areas and brought under strict control in most others.

People lost interest in the gypsy moth control project when it had seemingly done its job. In 1900 a special investigating committee decided that further control work was unnecessary. In fact this committee argued that the moth did not need to be thought of as a serious pest.

The views expressed by this committee, plus the lack of public concern now that the moth was under control, led the legislature to make no appropriation. Without money the control work by the state was brought to a stop.

Unfortunately, the work of the gypsy moth was not to come to a similar halt.

During 1900 and 1901, there was very little serious damage by the gypsy that called public attention to it. However, a trained observer had little trouble finding evidence of its rapid multiplication in woodland and private estates that had become neglected. In 1902, many of these estates, particularly those with non-resident owners, were very badly damaged and a number of woodland colonies were swarming outward in all directions. By 1903, the gypsy moth was well established in serious numbers in many parts of its former range. In 1904, everyone realized that not only had this pest reinfested most of the areas from which it had been removed, but had extended its territory. The hungry caterpillars swarmed everywhere leaving a destruction in their path as thorough as if fire had swept the area.

It is hard for us today to imagine the hordes of these caterpillars that humped their way across the Massachusetts' landscape. A report of these invasions states that the caterpillars were everywhere. In one report a citizen stated: "The place simply teemed with them, and I used to fairly dread going down the street to the railroad station. It was like running a gauntlet. I used to turn up my coat collar and run down the middle of the street. One morning in particular I recall that I was completely covered with caterpillars, inside my coat as well as out."

To understand how this insect spread out across the landscape it is necessary first to take a look at its life history. From mid-July to mid-August, adults can be

A female gypsy moth laying eggs.

found emerging from their reddish brown pupae cases. The male gypsy moth has a rather slender body and ranges from tan or dark brown in color. The front wings have wavy black bands and arrowhead markings. Wingspread is about 1½ inches. The female gypsy is nearly white. She is much larger than the male with a wingspread of 2 to 2½ inches. Her abdomen is heavy and cylindrical, with a covering of buff-colored hairs.

Because of her heavy body, the female cannot fly, although males are strong fliers and are normally to be seen in zigzagging flight near the ground in search of mates. Since she is flightless, the female lays her eggs near where she emerged from the pupa. This may be a tree trunk, under loose bark, under rocks, on stone walls, buildings, or in piles of rubbish.

Life for adult gypsy moths is short. Females mate within a day or two of emerging from the egg and die once the eggs are laid. The males likewise live only a very short time. As adults they cause no damage for they do not eat anything during this brief and final fling at love. They seem to have one purpose—reproduce the species.

The eggs await the far off spring to hatch. They require an exposure to cold temperatures to stimulate the development of the larvae growing within. Extraordinary cold can be a killer; however, it must be extreme cold, for the eggs can survive at least twenty degrees below zero. Many of the eggs are insulated by snow, while getting the necessary chilling.

A period of warm weather stimulates hatching in early spring. Late April or early May is the usual time. If cold wet weather follows the hatching, the larvae don't move far from the egg cluster, but as soon as the weather is right they spread out in search of food. The newly hatched caterpillars can normally survive about a week without feeding. Before their first skin shedding the caterpillars move very slowly and are active only when the temperature is above 60°F.

When disturbed during their feeding, the hairy young caterpillars drop from the leaves attached only by a silken thread. The insects usually start feeding at dusk and continue throughout the night. At daybreak they lower themselves on silken threads to find a shelter for the day. It is at the descent that they can most easily be carried some distance on the wind. Depending upon the air currents, they have been known to rise as high as 2,000 feet and to be carried distances of 20 miles. That's quite a ride for an animal that size.

The caterpillars feed heavily and grow very rapidly. Naturally the bigger they grow the more they eat each day. From hatching to the final molt they will have grown from 1/16 inch to 1½ to 2½ inches in length. The period of growth from one skin shedding or molt to another is called an *instar.* Caterpillars destined to become adult males go through five instars while females go through an additional instar.

It is estimated that 70 percent of all the leaves they eat is done during the last instar which generally occurs around the middle of June. This is when we see the terrible effects of their defoliation, and it will continue through the end of June or early July when they reach full growth and stop eating. Then they begin their quest for a protected place to form their pupas. In the ten days to two weeks they spend in their pupal chambers, their bodies go through an almost magical transformation to emerge as adult moths. Since not all eggs hatch at the same time, there is a brief time in early July when you may find all four stages of the gypsy moth at one time.

The egg and caterpillar stages permit the spread of the insect. Since the eggs are laid on any nearby surface, in heavily infested areas this may be boxes or crates, freight cars, automobiles, plant nursery stock, lumber or a host of similar materials. If these items are transported to other areas the eggs go with them, soon to hatch and to start their cycle of destruction. We have already discussed how wind carries some caterpillars around. In addition, grown caterpillars, searching for new food supplies or a place to pupate, may end up in autos or trains and be accidentally carried to new regions. In fact, the automobile was one of the major agents for spreading the insect around the state and in time across the country.

By such means the gypsy moth had re-established itself in Massachusetts in the four years since control measures had been brought to a halt. The report suggesting that the stories of the damage from the caterpillars were greatly exaggerated was not believable. New legislation to fight the gypsy moth was signed by Massachusetts Governor Douglas on May 8, 1905. It provided for cooperative efforts by the state, towns and individual landowners in bringing the insects under control and, if possible, eradicating them completely.

The head of the gypsy moth caterpillar has yellow markings while the body is dark grayish brown finely peppered with darker spots.

It was not long, however, before the insect began to appear in the adjacent New England states and New York. In the early invasions the authorities in these states didn't get any more excited than the authorities did to Professor Trouvelot's warnings back in the 1860's. This gave the gypsy moths ample time to multiply.

The battle followed essentially the same plan as earlier. Eggs were creosoted, shade trees burlapped, and spraying was carried out. Some areas were burned over as soon as the caterpillars hatched. In many areas brush and overhanging limbs were removed from roadways so that caterpillars would not be accidentally transported by dropping onto moving cars. A major innovation was to bring beetles, wasps, and flies in from Europe and Asia. These insects were parasites of the gypsy moth and helped to check moth numbers in those lands.

But such actions were not sufficient. With the help of modern transportation the gypsy was again on the move. As it traveled across state borders, it became an interstate menace. In 1906, the Federal government entered the scene and enacted quarantine laws to halt the spread of this pest. Nonetheless, pockets of the insect began cropping up in western New York, Ohio, New Jersey and western Pennsylvania. Although most of these outbreaks were quickly eliminated, one outbreak in New Jersey in 1920 took 15 years to eradicate.

In New England the gypsy moth wandered on, and by 1922 it had begun to move into eastern New York. This problem was a drastic one and called for drastic measures. A battle line was drawn, and intensive control measures were initiated to prevent further westward spread. This was a zone 25 miles wide and 250 miles long. It was just east of the Hudson River and ran from Long Island to the Canadian border. It was a good idea but despite a great deal of effort it failed to permanently stop the wanderings of the gypsy. The suspicion is that young larvae were blown across the line by winds.

Pennsylvania discovered an infestation near Pittston in July of 1932. Investigations revealed that the insect had already spread over 400 square miles. More infestations were found in central New York, New Jersey and Michigan. Every effort was made to fight these new pockets of moths, because it was feared that they would soon spread to every forested state. The great oak forests of the Appalachian Mountains, the Midwest and the South were particularly in peril. The appetite of the gypsy is partial to most trees and shrubs throughout the region.

Wherever the gypsy went and developed it set in motion many changes. Not only did such infestation destroy valuable trees, and the scenic value of tourist areas, it was also a public nuisance to people living in the areas. It removed some trees allowing other species to gain an advantage. In some regions it was able to change the general makeup of the forests. Conifer trees were often completely eliminated and in several forests maples were left standing and the oaks were destroyed.

Despite the damage, control measures available at the time were just not effective enough. Great hopes were held for its control by imported natural enemies. By 1934, over 40 different predators and parasites had been brought in.

But only nine parasites and two predators out of this number had become well established.

In 1944, the promise of a permanent answer arose. At that time the War Department gave the state of Pennsylvania about 100 pounds of DDT to try out on the gypsy moth. The tests were widespread and they showed over a four year period that this poison really did the job. DDT promised not only gypsy moth control but possibly eradication. By widespread spraying of this material from airplanes it was hoped we could get rid of this creature once and for all. These attacks were successful in practically the entire area treated. It looked as if the problem would soon be solved. But now a different problem lurked around the corner. There was growing evidence that DDT was a very dangerous chemical to be turning loose on the environment. The evidence at that time was far from clear but it was enough to cause some citizens to object strenuously to the broadcast spraying of the chemical from aircraft. New York citizens got the program halted through court action. In time the court ruled against the citizens and the spraying was started again. In the meantime, the gypsy moths munched on. But the spray program never really got going full swing again and finally came to a standstill. Evidence against DDT continued to grow, and today it is well recognized that massive use of the poison cannot be tolerated since it remains in the environment for many, many years. Each new spraying is added to the previous one to build up dangerous amounts of the chemical. Furthermore, it builds up in normal food chains to threaten creatures other than those for which it was intended.

Poisons, such as Sevin, have been tried with mixed success. It would seem that just when we thought we had a real chance to eliminate the gypsy moth once and for all, we failed again. The price in damage to the environment was too high, higher even than the destruction caused by the insect. What direction the efforts against the gypsy moth will take is not yet clear. Choices seem to include: either better means of biological chemical control or use of the short-range programs started before DDT.

In any case it seems that the results of Professor Trouvelot's accidental release are going to be with us for many more years. The battles to contain the spread of this insect and to control its local rise in numbers will go on. It promises to be a long and difficult war.

THIRTEEN

With Sting of Fire

THE IMPORTED FIRE ANT

The imported fire ant came quietly to this country from South America. No one is even sure when it arrived. It was first reported from Mobile, Alabama, in 1918. Undoubtedly it had already been with us for several years, since the species is easily confused with our own native fire ant. It probably slipped in among cargo from South America.

The imported fire ant, or IFA as we shall call it, is a medium-sized, blackish ant with an orange-brown band at the base of its abdomen. For about ten years the IFA was found only in or around the city limits of Mobile. It spread its colonies quite slowly, expanding outward only about a mile a year.

Nobody paid much attention to the new arrival. It was true that it had a very bad sting but so did its native relative. Some people are highly allergic to the sting just as they are to bees, wasps, and hornets'. To most people, however, the sting was only a brief annoyance.

There are several color phases of the IFA. The dark phase was the first to arrive. Then around 1930, a paler and smaller phase arrived. Its color is basically a pale red.

The pale variety spread more quickly into the countryside. It pushed outward at a rate of about three miles a year. Later other factors helped it to spread, and by 1963 the IFA was scattered over 31 million acres in nine southern states. By 1970, it was to be found over 120 million acres from eastern Texas and Oklahoma east to North Carolina.

Few insects that have been introduced have stirred up the controversy that the IFA has. To some people the species is a monster marching across the land that must be eliminated from this continent at all costs. To others it is a potentially useful predator on other insects. They see it as having only a few drawbacks. At worst these people consider the IFA a minor pest.

Those who claim that it is a dangerous invader have enlisted the aid of a number of powerful supporters in government and industry. Vast sums of money have already been spent to eliminate the IFA from our country. In 1969, a 12 year plan was drawn up as a final assault on this ant. Carrying out this plan would cost $200 million!

Those who are opposed to such a campaign to eliminate the IFA point out that the poisons used are far more destructive than the ant. Furthermore, they say the ant does not do the damage it is accused of. In fact, the IFA may even be a beneficial insect.

With such large sums of money involved, the battle lines are drawn. Emotion is more common than reason among the participants. The winners of the contest are yet to be determined.

Down in Argentina, one of the main homes of the IFA, the insect is not considered an economic pest. Its food is made up mainly of insects, and most Argentine agricultural specialists consider the IFA to be beneficial.

In the United States a very different picture has been painted by those who want to get rid of the fire ant. The United States Department of Agriculture has said the following about the IFA in two of its publications.

"This ant damages vegetable crops by sucking juices from the stems of plants and by gnawing holes in roots, stalks, buds, ears, and pods. It injures pasture grasses, cereal and forage crops, nursery stock, and fruit trees.

"When their mound is disturbed, these ants attack by sinking powerful jaws into the skin, then repeatedly thrust their poisonous stingers into the flesh. Fire ants may attack and kill newborn pigs, calves, sheep, and other animals; newly hatched chicks; and the young of ground nesting birds."

Unfortunately, such statements are not supported by sound research. At best they contain some half-truths. Such statements do arouse people against the fire ant and that is their main task.

In 1957, S. B. and K. L. Hays undertook experiments to find out just what the IFA's food habits were in this country. They observed the ants collecting food in the field. They pulled apart their nests to find what food was stored inside. Back in the laboratory they offered ants a variety of foods to see what they would choose.

What did they find out? They found that fly larvae were a favorite food. They found many other insects as well,

To find out what damage the imported fire ant does, scientists have opened ant mounds to discover what food was stored inside.

such creatures as: termites, weevils, beetles, cutworms, and aphids. They also found snails and spiders. Apparently the IFA's also eat each other from time to time. In the laboratory they tested the IFA's preferences in plant foods. Of many varieties of seeds offered only peanuts, okra, and corn were eaten. The others were not touched. Eighteen kinds of seedling plants were transplanted into IFA mounds. They were pulled up and examined carefully six weeks later. They found no damage. Eighteen species of plants were grown from seeds planted in the mounds. There was no damage to the seeds and of the resulting plants only okra was eaten.

This research and other similar studies indicate that the imported fire ant is primarily an insect eater in this country just as it is in South America. It is a predator of many insects including some agricultural pests.

Plant material is a small part of the diet and is largely limited to plants with a high oil content.

But what about the stories of the killing of baby mammals and ground nesting birds? Quail in particular are said to be victims of the fire ant.

Apparently the stories of the ants' killing newborn calves and pigs are fiction in the same category with milk snakes that milk cows or hoop snakes that take tail in mouth and roll away. That is, they are stories built by imagining what went on before a particular observation. Ants seen crawling on a dead calf or pig are thought to have killed it. In truth they had only discovered it in their search for food. They will eat dead or dying animals if they find them.

The danger to nesting quail was an argument sure to win sportsmen to the side of those who wanted to rid the South of the IFA. The ants were accused of entering the egg as soon as the chick opened a hole in it. The ants were then supposed to kill and eat the chick before it ever freed itself from the shell.

In Alabama, the Cooperative Wildlife Research Unit at Auburn University looked into these charges. Their research revealed that imported fire ants rarely attacked and killed normal hatching quail chicks. Most chicks that were attacked were not normal and healthy. In many cases, the ants were attracted to nests after the chicks had already died of other causes. Droughts often affect the hatching process in several ways, preventing chicks from escaping the egg after they have broken through. Ants attracted to such dead or dying chicks may also attack chicks that are still living.

The research also showed that the adult birds can keep limited numbers of the ants out of the nests. Deserted nests, with some chicks surviving to peck through the shell, were the most likely to be attacked.

The general conclusion was that the limited destruction of quail by the imported fire ant had no significant effect on the total quail population. Certainly the effects of some poisonous sprays used to fight the ants were far more damaging to quail populations than the ants themselves.

The charges of extensive damage to crops and wildlife may well have been developed to get rid of the ant because of its burning sting. Both the native and imported fire ant possess a sting that can in no way be considered a fun experience. People working in fields infested with fire ants may well get stung if they step in an ant hill. The sting usually brings a moment of searing hot pain followed by pimple-like pustules that last three to ten days.

In a bulletin from the Alabama Agricultural Experiment Station, G. H. Blake says, "If you've never had a bout with stinging ants consider yourself lucky. When the insect involved is the imported fire ant, the bout generally becomes a rout—and in short order! Imported fire ants are vicious stingers and attack without provocation." By contrast, D. W. Coon and R. R. Fleet writing in Environment magazine state: "According to entomologists working with the IFA, the foraging workers do not attack when approached. Instead, they communicate, presumably via chemical substances and retreat to the nearest entrance of their mound." And a study from Argentina reported: "A large proportion of the agricultural produce of Argentina is harvested by hand. Persons working in citrus and apple orchards and cotton fields reported being stung only when they stood on a mound."

Again we find conflicting statements on the seriousness of the menace that the fire ants are supposed to be. One thing is certain about the sting of this insect: there are some people who are seriously allergic to it. These people need prompt medical attention if stung. Most people become allergic only after being stung several times before. The little white sores or pustules that rise near the sting may leave a permanent scar if scratched.

There is no doubt that the sting of this insect is not pleasant. On the other hand, we do not spend millions to

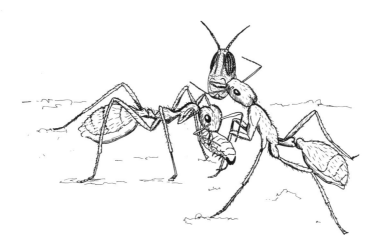

The imported fire ant (IFA) is an alien insect whose disastrous effects upon native wildlife is disputed.

get rid of all wasps, yellow jackets, bumblebees, hornets and honeybees. In fact, of the 460 deaths caused by poisonous animals in the United States in the years 1950 through 1959, only four were caused by ants of any species. In the same period bees caused 124 deaths, wasps 69, yellow jackets 22, and hornets 10.

The other major charge leveled against the imported fire ant is that farm machinery is damaged or broken when it strikes the ant mounds. In some areas, and with some machinery, this is quite true. However, clogging of machinery with dirt and an occasional bent rotary blade must fall more in the class of nuisance than extreme economic hardship. Back in Argentina this was no problem since farmers don't clip or mow their pastures. On cultivated land they had little trouble for the ants cannot tolerate cultivation and frequent disturbance of their mounds.

The evidence seems to be that the fire ant is a nuisance. Its sting can be irritating and occasionally dangerous. Its mounds sometimes interfere with modern mechanized farming. On the other hand, it does feed on a variety of insect pests and feeds only occasionally on plants and then only a select few. Its damage to wildlife and livestock has been wildly exaggerated.

Local controls in some areas of human activities may well be called for. In other areas the insect may actually benefit crops as a natural predator. However, at the moment, the imported fire ant is still the target of a massive government sponsored campaign to totally eliminate it from North America. To date the work toward this end has not done much to push back the ant but it has had disastrous effects upon other wildlife.

The super insecticides—dieldrin, chlordane and heptachlor—have been widely used against the IFA. These chemicals break down very slowly and stay on the area for a very long time. Each new dose adds to the one before, building up very high amounts of the poison in an area. In addition it takes only small amounts of these poisons to kill other animals. As the chemicals were sprayed from airplanes over thousands of acres, widespread damage to native wildlife resulted.

The fire ant has been accused of killing quail. Yet on one study area sprayed with the super insecticides, 13 whole coveys of quail that lived on the treated area were killed. Two other coveys that ranged mostly off the treated area survived. As far as quail were concerned, if fire ants were the sickness, then the cure was far worse. In another study area meadowlarks and red-winged blackbirds were totally destroyed. Earthworms, chief food of the woodcock, were still loaded with the poison five months after the treatment. Not only did the poisons kill vast quantities of wildlife, it did not completely control the ants as predicted. In a year or so the ants were back.

After the publishing of Rachel Carson's highly important and world shaking book *Silent Spring,* the United States Department of Agriculture reconsidered its approach and stopped using these sprays. They then began a search for a new miracle insecticide that would kill the ant without doing damage to other wildlife.

The chemical they came up with was called Mirex. The use of the substance was quite clever. Studies showed that the main food item was insects and that plant material was used only as a last resort. (This was interesting considering the charges against the insect as a plant eating pest.) Further studies showed that the best baits to attract the IFA were those high in protein or fatty acid. They were cottonseed oil, soybean oil, peanut oil, lard and fancy tallow. Soybean oil was chosen; dissolved Mirex and oil were then absorbed by ground up corncobs.

The bait was spread out over the area to be treated. Foraging worker ants brought the bait to their mounds. According to the protocol of the ant society the queen is fed first and then the larvae. The poison is fairly slow acting but before long the whole colony is dead—queen, youngsters, and workers.

It sounds good but unfortunately the poison has not been able to control the fire ant for any length of time. This is true even after six applications of the poison in some areas. Furthermore, Mirex does affect wildlife. It reduces the fertility and hatchability of quail eggs and it has effects on the reproduction of mammals, too. It is also known to cause cancer in mice. Like others of its family of chlorinated hydrocarbon poisons, it is passed along in food chains. It increases in amount with every link in the chain. The more the poison is studied the more dangerous it appears. The Secretary of the Interior has sharply restricted its use on lands under his supervision. Mirex has failed to control the fire ant so far and continues to pose threats to wildlife and man.

We are faced with some important questions:

Is the fire ant really as dangerous as alleged?

Are the poisons used against the ant more dangerous than the insect?

Should we invest $200 million over a dozen years or so to try and rid ourselves of this insect?

FOURTEEN

Bringing in the Troops

VEDALIAS AND OTHERS

The citrus growers of California were in deep despair. Some were even pulling up their trees and selling out. In many of the citrus groves the fruit crop was a total loss. While in some of the groves, the trees themselves were killed. Their enemy was a small and inactive insect that fed on the sap of the leaves and twigs of the citrus trees. It was known as the cottony-cushion scale insect.

It was 15 years since the insect had first appeared in the groves in 1872. The cottony-cushion scale had spread throughout the whole California citrus growing region and it threatened to destroy the industry. During those 15 years no method of controlling the insect had been found. There didn't seem to be much hope that a control method would be found either.

Entomologist C. V. Riley, from the United States Department of Agriculture, became interested in the problem. He felt there were some clues to a possible solution. If poison sprays would not work perhaps natural insect enemies would. It was known that the cottony-cushion scale came from Australia. It had probably come to this country on nursery stock. It was also known that there was a parasite in Australia that attacked the scale and kept its numbers in check. In Australia the cottony-cushion scale was not a serious pest.

Riley felt that if the parasite could be located and brought to this country it might survive and feed on the scale insect. He decided to send a trained entomologist to Australia to obtain the parasite. He chose a young man who had been studying the scale insect in California and ways of controlling it. In fact, it was this man who had figured out that the insect must have come from Australia. His name was Albert Koebele.

Koebele arrived in Australia in late September of 1888. With the help of an Australian entomologist he quickly set out to find the parasite. It was a tiny fly whose name, *Cryptochaetum iceryae,* was far longer than the insect itself. The fly was quickly found and Koebele immediately sent many thousands of them back to California.

He was lucky to have completed his mission so easily, and more good luck still awaited him. As he probed around in Australia he discovered the previously unknown vedalia beetle—a relative of some of our native ladybird beetles. Furthermore, he found that both the adults and the larvae of this beetle fed greedily on the eggs and the larvae of the cottony-cushion scale. In fact they fed only on the scale. He quickly sent a shipment of the beetles back to California. They arrived November 30, 1888. Other shipments followed and by the end of March of 1889, a total of 514 had arrived. Koebele's luck held. The beetles thrived and multiplied.

The beetles were released into the citrus groves and within two years the scale was under complete control throughout the citrus growing region. Despair was put aside. The parasitic fly also became established and abundant but it was the vedalia beetle that did the main job of destroying the cottony-cushion scale insect.

Koebele's whole trip cost about $5,000, but his discoveries saved the citrus industry millions of dollars; in fact, they probably saved the industry. In addition it established Koebele as a pioneer in the new field of biological control, that is, the use of natural enemies rather than chemicals on insect pests. The $5,000 was very well spent.

Most of our destructive insect pests are introduced from other parts of the world. Back in their native lands they are usually kept under control by a variety of diseases, predators and parasites. These enemies do not completely destroy the insect, but they do keep its numbers in check. When the insect comes to a new land it usually comes without its enemies. Often the new land does not have enemies to take their place. When that is the case there is nothing to keep the insect's numbers in check. The species multiply rapidly and if they eat our crops or trees they become pests. This is why it is mainly foreign insects that become pests in our country. For the same reason, insects we would not consider pests may become so in another country. The idea behind biological control is to import the insect's enemies to help bring its numbers under control.

In 1893, Albert Koebele was appointed to the Hawaiian Territorial Board of Agriculture and Forestry. Over the next ten years he was responsible for bringing in and establishing more than 18 beneficial insects on the various islands of Hawaii. These insects came from Australia and Asia. The work was very successful, and as a result the important sugar cane crop of Hawaii is free from serious attack by any insect pests.

One of the main reasons for Hawaii's long and continued success with biological controls is its climate. The parasites can develop at any season. Since there is no cold winter there is no hibernation period for the insect pests. In other climates the parasites have to find another food species while their main target is hibernating. If they don't they die. Neither does Hawaii have long periods of hot dry weather which causes the natural enemies to become inactive in summer. Such weather conditions have prevented the use of promising predators and parasites on the United States mainland.

Largely due to the successes of the pioneering Koebele, a number of entomologists began to study the biological control of introduced insect pests. And 50 years after Koebele imported the vedalia beetles, a different insect was brought into this country to control a plant, rather than an insect.

Around 1900 a pesty weed showed up in California in the Klamath Valley. Thus Californians call it Klamath weed. It is a pest not only because it makes grazing livestock ill, but because it crowds out desirable range plants. By 1952, California alone had some 400,000 acres infested with it. The cattle and sheep that feed on the Klamath weed seldom die, but they become very irritable; they become hard to corral and sometimes the only way they can be handled is to corral them and feed them other food. In most cases, after a day or two the effects of eating the weed wear off.

In Europe and much of the eastern United States this same weed is called St. John's-wort. This is because legend has it that it blooms on June 24, the day of St. John the Baptist. Scientists call it *Hypericum perforatum* because the leaves have tiny pin prick size perforations or holes in them. These can be seen if you hold a leaf up to the light.

St. John's-wort is a perennial weed; that is, once it gets established it comes up again year after year from the same root system. Perennial weeds are hard to control. Chemicals such as borax and 2-4D were tried but without much success. They were expensive but, perhaps of greater importance, the weed is spread over large areas. Many thousands of these acres are not easy to reach with spray equipment.

If chemicals wouldn't work perhaps there was another way. Many insects eat plants, much to our distress. Perhaps, however, there was an insect that fed only on this weed. The insects could be put to our advantage in controlling a weed.

The Australians made the first moves in this line of weed control. In 1920, they started looking in Europe for enemies of St. John's-wort, insects which would eat the weed but not their crops. All insects were tested on 42 different plants representing 19 plant families. Two kinds of *Chrysolina* beetles were finally chosen for release on the Australian ranges. Both fed on St. John's-wort leaves.

After the beetles and their offspring had been at work for about eight years, they were given a good report card by the Australian officials. American scientists had been watching the Australian experiments carefully. Shortly thereafter biologists at the University of California were given permission by the United States Department of Agriculture to start similar tests to be sure these insects would not eat our crops. They decided to try three insects. Two were leaf feeders, one was a root borer. None have common names. The two leaf feeders are called by scientists *Chrysolina hyperici* and *Chrysolina gamellata.* The root borer is named *Agrilus hyperici.*

The time was now the 1940's and there was a World War raging in Europe. This made it impossible to collect insects there. Fortunately the Australians offered a helping hand with those they had. The United States Army Air Transport Command flew the shipments in.

The two leaf feeding beetles passed the tests in the mid 1940's, but work with the root borer ran into difficulties and had to await the end of the war. Both of the leaf eating species were well established by 1948, and we did not have to import any more. The number of insects grew astoundingly and they were quickly introduced into Idaho, Washington and Montana. *Chrysolina gamellata* has been the more successful of the two species. It begins reproducing quickly when the fall rains come. At this time the weed sends out strong leafy shoots at the base of the flowering stalks. The beetles lay their eggs on this leafy growth. The young larvae grow during the winter months, and by midwinter and early spring the larvae are between a half

The Chrysolina beetles (left) and the vedalia beetles (right) were brought to America by man to naturally control weed and insect pests. The vedalia beetles are shown eating cottony-cushion scale.

grown and fully mature state. At this stage they really destroy the St. John's-wort leaves.

When the larvae are fat and fully mature they dig into the ground and become pupae. In April and early May the adult beetles emerge on the leaves and blossoms.

By late June and early July the beetles are well fed and prepared for their summer sleep through the dry season. They spend this time beneath small stones, in cracks in the soil, and under a variety of other objects. The plant also becomes inactive during this dry period. The seeds ripen then but the plant drops its leaves and the branches and twigs become hard and woody.

Unfortunately *Chrysolina hyperici* is not as active as its cousin in the moist conditions of the fall. This means it is likely to lay its eggs late. Should this happen the insects may still be in the pupal stage when the dry season arrives. If they are, they usually die. Consequently *Chrysolina hyperici* has done very well only on limited areas when conditions are right.

The Klamath weed is mainly a pest in areas where the soil is moist from winter to early summer and then dry later in the year. Overgrazing helps it spread by removing competing plants. Once the Klamath weed gets started, the other plants can't come back. The beetles not only destroy the weed in many cases, they seriously weaken those that are left. When the land is properly grazed by cattle, the range plants can come back and recreate conditions that help prevent return of the weed.

Chrysolina beetles as weed controls are now a permanent part of America's wildlife. They are making a solid contribution to reducing the Klamath weed or St. John's-wort in many areas. Where man cooperates with the beetles and uses the once weedy areas wisely, the range improves and fewer cattle are sickened by eating the weed.

Unfortunately farmers, particularly those for whom farming is a big industry, have often been impatient with biological control. They want insect and plant pests eliminated, not controlled. In a number of cases this has been possible largely through a combination of methods.

In the 1940's, due to war related research, the first of the super insecticides, DDT, was created. Since 1945, it has been widely used along with a whole series of new compounds, many even more poisonous than DDT. They worked much faster than biological controls and usually did a more thorough job at first.

By the 1960's, it was revealed that these powerful poisons were not the whole answer. A number of pest insects developed a resistance to them. This stimulated development of even more potent poisons to meet the challenge. At the same time these poisons were poisoning many other creatures for whom they were not intended—beneficial insects, wildlife and even man.

In many cases the sprays destroyed the introduced parasites and predators but failed to completely control the introduced pest. For example, DDT was used in the California orange orchards to control the citricola scale. The spraying was soon followed by the first outbreaks of cottony-cushion scale since 1890. The poison didn't have much effect on cottony-cushion scale, but it did destroy the valuable vedalia beetles.

During the 1950's and 60's, vast tonnages of pesticides were dumped helter skelter all over the world landscape. Someday this may be known as the Great Pesticide Period. It may be many years before we truly know how disastrous the broadcast use of these poisons has been to Earth as a whole. As we mentioned before, the tide was turned in 1962, with the publication of Rachel Carson's *Silent Spring.* But it may take us years to truly reduce the use of pesticides to any reasonable level.

In the meantime, although pesticides have destroyed some of the work done by Koebele and those who followed him, there is a new appreciation for biological control. More money is going into the search for ways to fight insect pests.

In the years ahead we will owe much to the various methods of biological control which fight the uninvited guests from other lands.

MAN AND

OTHER ALIENS

"The white man is an imported weed." So wrote Henry David Thoreau in his Journals. He was only partly right. All men were imported weeds to this continent. The first human being appeared in North America over ten thousand years ago. They came on foot across the then existing land bridge between Alaska and Asia. They wandered down the open corridor between the ice sheets that covered much of North America.

How many waves of human wanderers ventured down the broad pathway we will probably never know. But enough came to establish man in North America. The species flourished and eventually spread across both continents. As many local cultures developed, some were more in harmony with the land than others. Early man in America left his mark on the land and the wildlife. There is growing evidence that he helped some of the great Ice Age birds and mammals to an earlier extinction than would otherwise have been expected. And fire was deliberately used by these early Americans to change the landscape.

When the white man came by boat, Columbus was widely credited with "discovering" America. There is much to suggest that he was not the first to discover America but only the best publicized. People from much earlier western cultures may well have landed on the American shores and established small colonies.

The invasion of North America by the white European man after 1492 has parallels with the introduction of the Norway rat. Just as that animal pushed out its earlier arriving relative the black rat, so the white man pushed his Indian relatives off the land. Yet, whatever disruption of the land and wildlife these first Indians brought about, it was nothing compared to that of the European invaders.

The Europeans brought with them the tools and culture they had developed over several thousand years. It was a highly technical society. With these skills they cleared the forest, plowed the grasslands, mined the earth, built cities and towns and factories that polluted the air and the water, and invented the automobile and airplane. Many such activities greatly changed the face of North America.

As a result some kinds of wildlife have become extinct —the passenger pigeon, the Carolina parakeet, the ivory-billed woodpecker, the Labrador duck and others. Many more species are on the brink of extinction or are seriously endangered. This is because the environment they need has been destroyed. In its place is a new man-created environment quite different from almost anything that existed before European man arrived.

Indeed there were some kinds of wildlife that found the man-changed environment more to their liking. When old forests are cut, new brushy growth comes up to replace them if the land is not farmed or built upon. Deer need this kind of growth for food. There is more such food available now than before the white settlers came. And there are now more deer than before the settlers, in spite of heavy hunting by man.

Rabbits, mice, and other small rodents like the food and cover that farming activities provide. Foxes which eat such creatures have also multiplied. The blackbird tribe, grackles, redwings, cowbirds and others have also prospered, feeding heavily on the grains that man grows. They have done so well that today they nearly rival in numbers the unbelievable hordes of passenger pigeons of the past.

Man has not been content to let native animals adapt to the new habitats he has created. Hunting and fishing have long been a favorite occupation of men. As man's activities destroyed the native habitats many of the native game species became less common. The magnificent wild turkey disappeared as the forests disappeared. The ruffed grouse also had special forest needs that were easily disturbed. The prairie chickens disappeared as the prairies were plowed. To fill the demand for huntable birds, men looked to other parts of the world to find species more suitable for survival in the man-shaped environment of America. They tried many kinds, but only the ringnecked pheasant, the Hungarian partridge and the Chukar partridge so far have found a home here. In general they now occupy habitats not used by other native game species.

The farming and industrial activities of man have also destroyed the habitat of many of America's big game animals. The bison and the pronghorn and the elk were at home on the rich prairies and plains that now grow our grain or graze our cattle. Domestic sheep diseases have helped to reduce the numbers of wild mountain sheep and the domestic sheep have eaten food that mountain sheep previously ate. There are groups of people today who are trying to find big game animals from abroad that can find a home here. They are particularly looking for animals that could use the overgrazed grasslands and semi-desert lands that have grown up to woody shrubs.

Many of the people who came to America found the land wild and forbidding. They grew lonesome for some of the more beloved birds and animals of the land they had left and they also wanted familiar flowers and bushes. These desires led to attempts to bring these plants and animals to new land. Brought in the hopes that they would find a home, most species just could not survive.

Some found a home in a tiny corner of the land such as skylarks near Vancouver, spotbreasted orioles near Miami, European tree sparrows around St. Louis and the European goldfinch on Long Island. The pigeon found a home in man's cities but never became a truly wild species. Only in a very few places like the sea cliffs near Nahant, Massachusetts, and the rocks around Niagara Falls, have these birds taken to their natural nesting sites. Window ledges and decorative architectural features on man-made buildings are preferred. Man through his waste and carelessness also keeps them fed.

The desire for flowers and shrubs also brought some troublesome weeds and insect pests that came along in the soil or attached themselves to the plants. Most of our serious pests came to America this way. A few animals were imported to try and control these pests. The giant toad, Mediterranean geckos, and the praying mantis were brought here because they were general insect-eaters.

Many Americans like to keep pets. Guinea pigs, gerbils, golden hamsters, canaries, parakeets and tropical fish are popular. These are all creatures that do best in warm dry climates. Even though many get loose or are deliberately let loose each year, none have become well established in the wild except some of the tropical fish. The strange walking catfish is a prime example. Officials in the southwestern states, however, are very concerned about the gerbil. Some of their climate and habitat is potentially suitable for this animal. If it established itself it could become a serious pest.

The domestic animals of man which become wild also can cause damage to native wildlife and the habitat. The burro and the hog are good examples of this. Under some circumstances even the dog and cat can be bad for wildlife. A few have become completely wild, others lead double lives. Part of the time they live at home as loving, peaceful pets. The rest of the time some become marauders. Dogs may form packs and hunt down deer and other animals. Sometimes these packs even attack humans. Cats usually act individually, catching such wildlife as they can, small rodents and birds.

Thus, for many reasons and in many ways, man has shuffled the fauna of the world. Not just in America but wherever he has gone.

In this book we have told only about newcomers to America. But we could have told similar stories about our gray squirrel in England, our muskrat in Europe, the European rabbit in Australia, trout in North Africa, the great African snail in Asia and the Pacific Islands and the European red deer in New Zealand.

The fact that man has become a species distributed around the world, and ever on the move, means that a number of small animals and plants are likely to be accidentally moved from one place to another because of him. Since past experience has shown that many of these become serious pests in their new land, nations have developed elaborate procedures to prevent unwanted newcomers.

Today airplanes are thoroughly sprayed to kill insects that may be aboard. Many plants and plant materials that arrive by boat, plane, or truck must be very carefully checked by specialists before they can enter the country. Laws have been passed that forbid the importing of known pests such as the mongoose. Other laws allow animals into the country only by permit. To obtain such permits persons or organizations must carefully prove their ability to control the creature and to provide references of their good judgment. Most animals in this permit category must be kept quarantined for days, weeks or even months to insure that they are not carrying disease.

Such inspection and quarantine procedures have greatly reduced the unwelcome stowaway visitors to the United States and to the other countries which use them. Occasionally some creatures do get by.

It is usually because people break the law and bring in plant material without telling anyone.

Of real concern today is the deliberate introduction of alien animals. Some of them, particularly the large hooved animals, are in danger of extinction in their native foreign lands. There are people who feel we should try to bring some of these to America in the hope that they can survive here.

There are also people who want to bring in animals just because they like them. They can't go abroad so they want to see them wild somewhere in this country. Many sportsmen are constantly looking for a new game species. They feel that with all the hunters afield such game animals could never become pests. Fishermen search for fish that can survive in our increasingly warmer and more polluted waters. And last, but not least, the search goes on for the biological control of plants or animals that are already pests.

There are other individuals who are completely opposed to bringing in alien or exotic creatures of any kind. Some further suggest that we should rid ourselves of as many of these we now have as soon as possible.

Years ago only man's desires seemed to be considered. Very little thought was given to the kind of conditions the animal required. For this reason most of the many imported animals died. Those that were successful were often too successful. The animal became a pest.

It was indeed the success, or over success, of the house sparrow and starling that caused responsible leaders to introduce some controls on the importing of exotic animals for release in this country. Ecology is a fairly young science. It is the study of the interrelationships among plants and animals and the physical environment. It is the procedures and findings of this science that can help us predict the potential success or failure of an introduced species and its effect on the environment. Using this science we might have been able to prevent the disaster of bringing in the starling and the house sparrow.

It is to ecology that we must now turn before bringing in any new species. We can approach our investigation from one of two major routes. One is to study an animal in its native land in exhaustive detail. We can learn its food habits, reproductive habits, natural enemies, climatic needs, and effects on other species. Armed with this information we can examine the areas where we propose to release the animal. We want to find out what native animals it would possibly compete with or what human activities it would interfere with. What effect it might have on the vegetation and physical environment, and whether or not it would find conditions right for its own survival. Also important is whether or not there are native animals that could and would replace its natural enemies and keep its numbers in check.

The second approach is to examine a habitat to see if there is a possible unused role or niche in it for a new species. In the case of environments that have been drastically changed by human activity such as agricultural lands or overgrazed range that has become brushland, there may be a niche available. The task then is to define the niche as carefully as possible. What foods would be involved? What species are present to act as predators or parasites? What is the climate? With this information the world can be searched for a creature that is filling a familiar kind of niche elsewhere. Studies can then be made to see if that animal could come to America and fulfill a need here. Getting information for either of these routes takes a great deal of time and money. It is, however, the only wise thing to do if you want to bring in new species at a minimum risk to our native environment. Actually, such procedure should be followed in introducing animals from one part of our country to another. The western house finch has been brought east and is not particularly welcome. The armadillo of the southwest was taken to Florida to the displeasure of many Floridians. The southern race of bobwhites was used to stock habitats in the north. It didn't survive well. However these birds interbred with northern bobwhites and the offspring were less able to survive the winters than pure northern birds. This has resulted in a strong decline in the numbers of northern bobwhites.

In spite of intensive ecological studies of a species and of habitats, no one as yet can predict for certain just how well animal and habitat will fit. When an animal is moved to a new environment, no matter how similar, the animal must make some adjustment to survive. In the same way the environment will have to make some adjustments to survive the new species. It is a complex process.

For example, when the red deer found success in New Zealand it increased in great numbers for there were no predators but man. It overgrazed its habitat severely. The result was large scale erosion in the highlands and massive silting in the lowlands where the dirt was carried. This changed the habitat of animals and plants with which the deer did not directly compete.

To date, very few introductions of alien species have been made based on such careful studies as proposed. Most have been haphazard and resulted in general confusion. The state of Hawaii has suffered the most from the introduction of exotic animals. It has lost a large portion of its unique island plants and animals. In fact it has lost 60 percent of its land birds that were found nowhere in the world but Hawaii. Of the continental United States, Florida is the state with the most exotic species. Very few of them were brought in under government permits and none was released as a result of an ecological study.

Some introduced species are already creating problems. Man has changed the landscape drastically and his activities threaten to dry up the Everglades. His pesticides are destroying native species and he is destroying various habitats in many other ways. The red whiskered bulbul is a bird that is becoming a fruit-eating pest and possibly a threat to Florida agriculture. New laws now ban any further imports of the bird except under federal permits. The giant toad is replacing the native southern toad in and around cities in south Florida, while the Cuban tree frog is feeding on native green tree frogs and squirrel tree frogs. The mosquito fish, a native, is preyed upon by the introduced gar-pike top minnow. The black bass, a favored game fish, faces the competition of two introduced fishes, the African cichlid and the peacock cichlid.

One of the most spectacular introduced species of birds is the scarlet ibis. This large wading bird lives up to its name. It is a brilliant scarlet red. It has a native cousin that is white. The two birds are interbreeding regularly. If this continues for very long the white ibis will entirely disappear. It is being suggested that even though it is beautiful, the scarlet ibis should have its legal protection removed and that it should be destroyed if possible.

Another quiet battle is going on at the expense of a seldom thought about creature, the Carolina chameleon or, more properly, Carolina anole. Several anole species from the West Indies have been brought to Florida, largely as the result of the pet trade. They are gradually replacing our own Carolina anole.

Just whether other introduced species are a positive or negative addition to Florida is not always clear. It depends to a certain extent on your point of view. One thing is clear. There are a tremendous number of introduced species there. There will have to be many major or minor ecological adjustments to accommodate them.

Other states will probably not face the problem of aliens that Florida has. This is good. Texas, however, along with some other southwestern states, is eager to introduce African big game animals. You may ask yourself whether any new exotics should be introduced at all. Perhaps. To do the job right will cost a great deal of money. Could that money be better spent for other purposes?

In 1966, two excellent biologists, Frank Craighead and Raymond Dasmann, made a study of the extent of introduced big game animals in America. The study was done for the U.S. Bureau of Land Management which published their findings. These two men proposed a series of guidelines for any future introduction of exotic big game. The guidelines are just as applicable to other aliens as well. They have proposed that we must first show a real need for the "to-be-invited" species. In other words, the animal should have a desirable ecological, economic and recreational impact.

Sound evidence must be presented that the new species will fill a vacant niche, that is, one not filled or suitable for a native species. The introduction of exotic species should be governed by the necessity for protecting native plants and animals and for preventing conflict with other existing or proposed land uses. If there is any doubt, don't import.

The two biologists strongly recommended a detailed study of the ecology of the animal in its native land coupled with detailed study of the area where it is proposed the animal is to be released. Disease interrelationships must be studied and imported animals quarantined to assure disease free stock. Since hybrid animals are often more vigorous than either parent, precaution against hybridization should be taken. We should reject animals that have close relatives in this country with which they could interbreed.

They further proposed that before any full scale release there be a smaller, test release in an enclosed area where the animal is carefully studied and its effects on the environment are assessed. And they stressed that before the animal is released we know adequate control methods to prevent over-population or excessive spreading.

Such guidelines can help steer a middle ground between irresponsible introduction and release of no new species. If you want to see exotic animals the best thing to do is to visit a good zoo. Some people want to get closer than that. They then buy unusual pets. They seldom care for them for long. Some people just let them go wild. This is generally cruel to the animal for it faces an unknown environment and usually a lingering death. If it succeeds in the wild it may prove disastrous to native animals as the walking catfish is proving. It is more humane to the pet and to native species to have the pet put to sleep rather than to release it. It is even more humane not to buy such pets to begin with.

The North American continent has a rich heritage of native wildlife. Although man, the exotic, has become a dominant animal on the landscape, he has the knowledge to provide some refuge for most of the animals that remain. In fact it is to his benefit to do so. Much of the pollution and other human activity that threatens wildlife threatens man as well.

The question that faces us all is do we have the will to protect what remains of our wildlife heritage? Will we clean up the environment? Will we bring in only new animals that pose no threats to native animals? Will we bring in new species that fill only vacant niches, primarily new niches created by past human errors?

Man has acquired life and death power over all animals including himself. He must now show himself, and all other creatures, that he has the wisdom to use this power in the best interest of all life.

Acknowledgements

Illustrations and quotations are reproduced courtesy of the following:

ILLUSTRATIONS

page 14: Gordon S. Smith from National Audubon Society
pages 24, 93: Jeanne White from National Audubon Society
page 25: Florida Game and Fresh Water Fish Commission
page 35: The Bettmann Archive, Inc.
pages 43, 85, 102-103: U.S. Department of Interior, Sport Fisheries and Wildlife
pages 62-63: A. D. Cruickshank from National Audubon Society
pages 65, 113, 121: Grant Heilman
pages 70-71: Hugh M. Halliday from National Audubon Society
pages 78-79, 89: Massachusetts Audubon Society
page 107: United Press International Photo
page 120: The American Museum of Natural History
pages 129, 133, 141: U.S. Department of Agriculture
pages 145, 155: Charles E. Roth

QUOTATION

pages 86-87: Peterson, Roger Tory, *Birds Over America,* reprinted by permission of Dodd, Mead & Company

Index

American Acclimatization
  Society, 82
Ant, see imported fire ant

Bees, 118, 122, 145
Bighorn sheep, 44-48
Biological control, 123,
  135, 149-157
  climate, 152
  Koebele, Albert, 150-152,
    157
Boll weevil, 119
Borden, Richard, 60
Brook trout, see trout
Brown trout, see trout
Burros, 38-48, *illus.*, 43
  as pests, 44-48, 163
  protection of, 42, 44, 47

Carp, 99-104, *illus.*,
  102-103
  description of, 101
  Cyprinus carpio, 99
  introduced, 100
  polluted waters, 101, 104

Carson, Rachel, 147, 157
Cattle egret, 59-65, *illus.*,
  62-63, 65
  arrival of, 59-61
  description of, 59
Chemical control, 136-137,
  146-148, 153, 156-157
Chrysolina beetles,
  153-156, *illus., 155*
Clarias batrachus, see
  walking catfish
Corbin, Austin, 26
Cottony-cushion scale
  insect, 149-151
  arrival of, 149-150
  control of, 149-151, 157
Coypu, see nutria
Craighead, Frank, 169-170

Dasmann, Raymond,
  169-170
DDT, see chemical control
Drury, William, 59, 60
Dutch elm disease, 122

Ecology, 166-170
English sparrow, see
 sparrow
European corn borer, 119,
 122
European elm bark beetle,
 122

Foxes, 12-13

Gypsy moth, 124-137,
 illus., *129, 133*
 control of, 84, 126-128,
  132-137
 escape of, 125
 import of, 124
 life cycle of, 128-132
 spread of, 127-128,
  131-132, 134-135
Gysel, Leslie, 21-22

Hen, heath, 37, 73
Hogs, 20-28, 163
 feral, 23, 27-28, *illus., 25*
 wild boars, 23, 26-28,
  illus., 24
Horse latitudes, 49
Horses, 48-53
 protection of, 52-53
 reintroduction of, 48-50
 wild herds, 50-53
House sparrow, see
 sparrow

Import guidelines, 166-167,
 169-171
Import laws, 94-95, 164
Imported fire ant, 138-148,
 illus., 141, 145
 arrival of, 138
 control of, 146
 controversy over,
  139-146
 description of, 138
 sting of, 138, 143-146
Insecticides, see chemical
 controls
International Society for
 the Protection of
 Mustangs and Burros,
 53

Japanese beetle, 86, 119,
 illus., 120, 121
Johnston, "Wild Horse
 Annie," 53

Klamath weed, 152-156
 control of, 153-156
Koebele, Albert, 150-152,
 157

Lacey Act, 94-95

Mather, Fred, 110, 111
McIlhenny, E. A., 15-16
Morgan, Allen, 59, 60

Muskrat, 15, 18
Mustang, see horses
Mute swan, 74-81, *illus.,
78-79*
  import of, 77
  swan-marks, 76

Nightingale, 55, 56
Nutria (coypu), 15-19,
  illus., 14
  description of, 15-17
  escape of, 15-16

Peterson, Roger Tory,
  86-88
Praying mantis, 123, 163

Rats
  arrival of, 29-33
  black, 29-34
  Norway (brown), 30-37,
    illus., 35
Riley, C. V., 149-150
Ringnecked pheasant,
  66-73, *illus., 70-71*
  enemies of, 72-73
  import of, 57, 68-69

St. John's-wort, see
  Klamath weed
Silent Spring, 147, 157
Sheep, bighorn, 44-48
Sparrow (English or
  house), 56, 89-95, 119,
  165-166, *illus., 93*
  import of, 90
  subspecies of, 93-94
Stackpole, Richard, 59, 60
Starling, 56, 82-89, 93, 94,
  119, 165-166, *illus., 85,
  89*
  description of, 83
  import of, 82-83
  roosts, 86-88, *illus., 89*
Swan, see mute swan
Swan-upping, 76, 77

Terns, 36-37
Trout
  brook, 110-112, 114, 115
  brown, 110-115, *illus.,
    113*
  wild, 112-114
Trouvelot, Leopold,
  124-126, 134

Vedalia beetle, 150-151,
  157, *illus., 155*

Walking catfish, 7, 105-109,
  163, 171, *illus., 107*
  Clarias batrachus, 105
  description of, 106-108
  eradication of, 105, 108
  import of, 107
  labyrinthines organ,
    109
Weaver finch, 90
Wildlife Management
  Institute, 57

About the Author

From boyhood, Charles E. "Chuck" Roth has been fascinated with living things and their world. This fascination has led him along the paths of naturalist, educator and writer.

Mr. Roth earned an M.S. degree in Conservation Education from Cornell University, and since 1961 he has been Director of Education for the Massachusetts Audubon Society. He also serves as a consultant in environmental education to the United States Office of Education.

Mr. Roth is also the author of the exciting and well received book, THE MOST DANGEROUS ANIMAL IN THE WORLD.